Idaho's Chinook Salmon:

The Journey of a LIFE Cycle!

by Deirdre Abrams

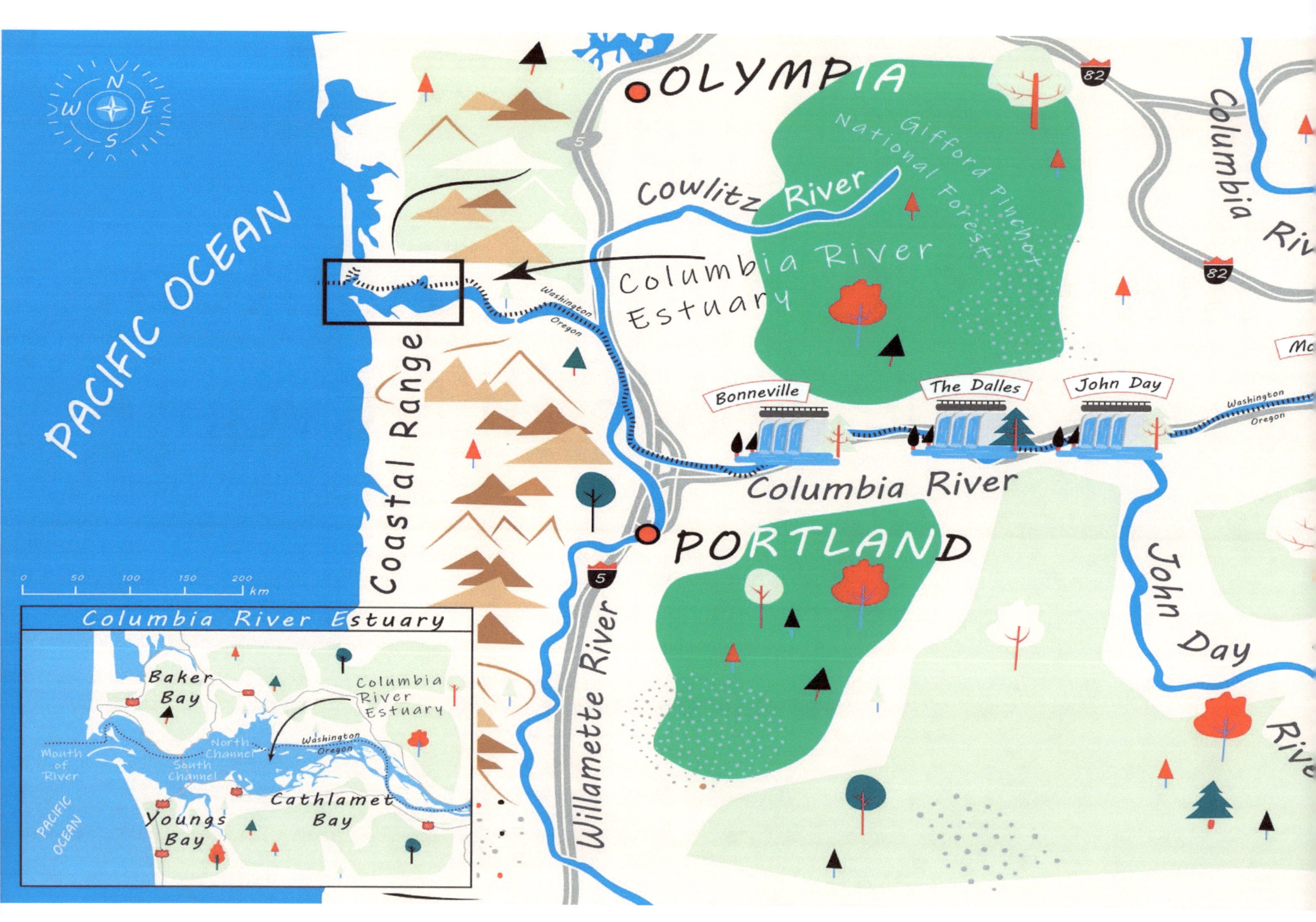

N
W E
S
PACIFIC OCEAN
OLYMPIA
Gifford Pinchot National Forest
Cowlitz River
Columbia River
Columbia River Estuary
Washington
Oregon
Bonneville
The Dalles
John Day
Washington
Oregon
Columbia River
Coastal Range
PORTLAND
Willamette River
John Day River
0 50 100 150 200 km
Columbia River Estuary
Baker Bay
Columbia River Estuary
Mouth of River
North Channel
South Channel
Washington
Oregon
Youngs Bay
Cathlamet Bay
PACIFIC OCEAN

90
Little Goose
Lower Monumental
Lower Granite
Ice Harbor
Clark Fork
90
Idaho
Montana
Lochsa River
Clearwater River
Lower Snake River
Washington
Idaho
Legend
State Boundary
Rivers
Interstate Highways
Cities
Washington
Oregon
Blue Mountains
84
84
Oregon
Idaho
Snake River
Salmon River
Salmon River
Oregon
Idaho

Paperback ISBN:978-0-578-77127-4

Library of Congress Control Number: 2020918998

First paperback edition December 2020

Printed by IngramSpark in the USA

Deirdre Abrams

McCall, Idaho

Front Cover Photo: FISHBIO

Back Cover: Artwork by Becky Bjork

Acknowledgements

A lot of smart, benevolent people were integral to the creation of this book. I am grateful for my husband, Jeff Abrams, who has always encouraged and believed in my creativity. He has worked as a fish biologist for Idaho Fish and Game, and his knowledge, compassion for, and dedication to save wild salmon has inspired me beyond words. Don Newberry, a retired fish biologist from the Boise National Forest, who has been like (a very young) dad to me, has volunteered in my classroom for about 10 years to help my students and me learn about and raise rainbow trout. He exudes a love of science and has been an inspiration to me. Craig Rabe, a fish biologist for the Nez Perce Tribal Fisheries (NPTF), has been an incredible resource. He has volunteered in my classroom to teach my students about salmonids for years, has enlightened me about salmon recovery, and has been a co-conspirator in serious excitement about salmon topics. John Gebhards, another fish biologist at NPTF, has volunteered for years to help my class with fish dissections, and his knowledge and playful nature has precipitated a love of science for so many students. Thanks to Bert Bowler for his extensive knowledge and all he does for fish! I thank the McCall-Donnelly School District for giving me the time to write and for allowing all of their teachers to have creative license. I thank Becky Bjork, a longtime friend and artist who painted the back-cover picture for this book! I am so grateful to FISHBIO for allowing use of their incredible photos. And, last, but not least, I must thank Becky Johnson, Scott Kellar, and everyone at NPTF for giving my students and me the opportunity to raise and foster awareness of Chinook salmon in our classroom.

Introduction

I have been fascinated by Chinook salmon for years. From the first moment that I saw wild salmon spawning on the South Fork of the Salmon River in Idaho, I instantly felt compelled to learn about and try to help to protect these strong willed, magnificent fish. That day, as I watched a female guard her fertilized eggs, literally from the minute they were laid in her redd to the minute she drifted off to die, I was struck by the power of her sense of duty and purpose.

Sadly, wild Chinook salmon, some migrating 900 miles to the ocean from the high mountains of Idaho, once made up nearly 50% of the Columbia River Basin's salmon runs. In 2020, and over the last twenty years, wild Chinook have been in crisis, listed as an endangered species, and face the growing threat of extinction. In fact, only 2 % of the Idaho's wild Chinook population remains.

I have had the pleasure of being a teacher for 25 years and counting. I feel so fortunate that the Nez Perce Tribal Fisheries is willing to work with me to educate students on this crisis by allowing us to raise hatchery fall-run Chinook in our classroom and learn everything about them through observation, biologist visits, and this book. Because Chinook salmon who originate in Idaho migrate farther than any other Chinook population in the lower 48 states and are endangered, the focus of this book is on them and their miraculous, obstacle-filled journey of a life cycle!

Contents

Chinook Salmon: Ocean Phase

Photo credit: NOAA Fisheries

Chinook Salmon: Freshwater Spawning Phase

Photo credit: Fishbio

Chinook: King of Salmon!

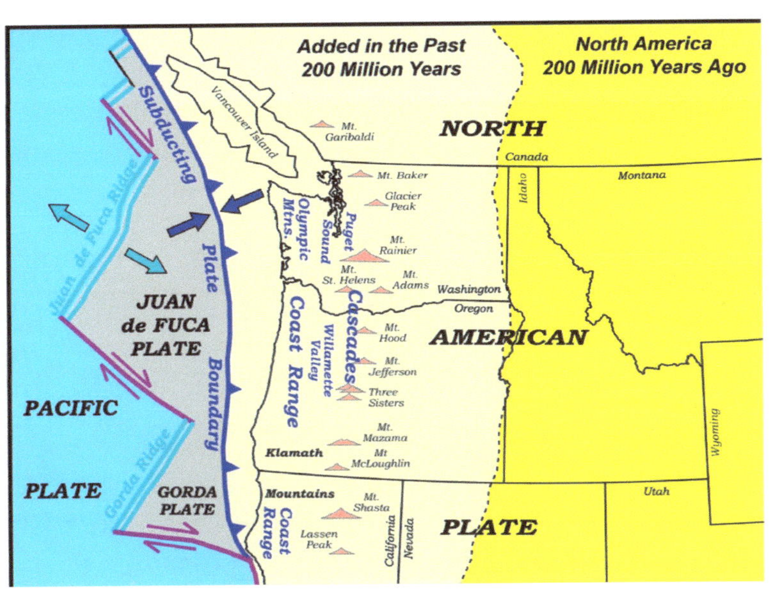

200 million years ago the coastline of the Pacific Northwest was near Idaho's western border. There was no Oregon and Washington!
Credit: US National Park Service

> "The extreme migration of the spring-summer Chinook salmon is one of the natural world's great journeys." ~Jim Robbins, NY Times

Long Distance Swimmers

Chinook salmon are **vertebrates**, which means they have a backbone like you. These cold-blooded fish were in Idaho when the first human beings stepped foot in what would become the state someday, about 12,000 to 16,000 years ago.

Before this, about 40-50 million years ago, salmon ancestors in this area mostly lived in cold-water lakes because the Pacific Ocean was much warmer than it is now. You would not have recognized Idaho at that time. The states that would become Washington, Oregon, and California were under water then, making Idaho beach-front property! About 25 million years ago, the oceans began to cool, and this is most likely the time that Idaho salmon began migrating to the sea. At that time millions of Chinook salmon existed in rivers all over this area, which later became Idaho, in the Clearwater River, the Salmon River, and the Snake River up to Shoshone Falls, as well as the many tributaries of all those rivers. Today, due to a variety of problems, Chinook are at risk of **extinction**. "It's hard to say, but now these fish have maybe four generations left before they are gone. Maybe 20 years," says Russ Thurow, a fisheries research scientist with the Forest Service's Rocky Mountain Research Station.

If there were an award for long distance, obstacle-filled swimming, it would go to the Chinook salmon born in Idaho. Although Chinook born in Alaska and Canada travel greater distances in miles, Chinook born in Idaho rivers swim the farthest with the highest elevation change in the world! Some of these heroic fish migrate from the freshwaters of the Salmon River in the Stanley Basin, around 900 miles and about 7,000 feet in elevation change. They must ride through rapids and over waterfalls, and through eight hydroelectric dams and reservoirs on the Columbia and Snake Rivers, all the way to the Pacific Ocean! "No other population in the world migrates that distance and that elevation, " says Bert Bowler, a former Idaho Fish & Game biologist.

Fish, such as salmon, who migrate to the ocean and back from their freshwater birthplace, are called **anadromous** fish. Unlike Atlantic salmon, Pacific salmon, like Chinook, are **semelparous**, meaning that they only migrate to the ocean and back just once to spawn and complete their life cycle.

Prior to the 1960s, it used to take Idaho-born Chinook a couple of weeks to travel the long distance to the Pacific Ocean from the Stanley Basin on a free-flowing Snake River. In contrast, now that they must navigate eight hydroelectric dams in the Columbia River Basin, it can take almost six weeks for them to make the same journey. This extra time consumes precious energy, as salmon do not eat while migrating. They must actually swim through the reservoirs upward of the dams, instead of saving energy by floating on the current.

Credit: U.S. Geological Survey

Biggest of Them All

Imagine finding that the fish on the end of your fishing line weighed 126 pounds! That might be more than the weight of the average person reading this book! This was the size of the biggest Chinook salmon ever caught. It was found in a fish trap near Petersburg, Alaska in 1949. Adult Chinook are the largest of all salmon and range in size from 24-36 inches yet may grow to be 58 inches in length. They are huge, powerful fish. Their average weight is 10-50 pounds, depending on where they live, and some have been known to weigh 130 pounds.

Mount of largest Chinook salmon ever caught- in a fish trap near Petersburg, Alaska in 1949. Credit: fishwithjd.com

What's in a Name?

The name for Chinook salmon refers to the Chinookan peoples of the Chinook Indian Nation, who live along the Lower and Middle Columbia River in Oregon and Washington. Because they are named after people, the word Chinook is capitalized, while other types of salmon are not. Take a look at the two-page map at the beginning of the book and find this area where the namesake people of Chinook salmon have lived.

Chinook salmon are also commonly known as King salmon, which is fitting because they are the largest species in the Pacific Salmon **genus**. The scientific name of this magnificent fish is Oncorhynchus tshawytscha (on-core-inkus, shaw-weet-sha). Other common names for Chinook are tyee, a Chinookan people's word meaning "the chief", hookbill, and blackmouth-because the inside of their mouths and gums are black.

The other salmon in the Pacific Salmon genus are chum, coho, pink, and sockeye. Steelhead, who are actually migrating rainbow trout, are often included in this list because they are also anadromous fish. Chinook, sockeye, and steelhead all migrate to the ocean and back from Idaho. Although the Nez Perce Tribe has reintroduced coho to the area, the last wild, solitary coho returned to the Snake River in 1986.

The Pacific Ocean is home to five species or kinds of salmon – chum, sockeye, Chinook, coho and pink. Idaho once had three species that lived here – coho, Chinook and sockeye. Coho are now gone from Idaho, but we still have Chinook and sockeye.

Here is an easy and fun way to learn and remember the names of the five salmon that live in the Pacific Ocean – use your hand as a guide! The thumb stands for chum. Use your pointer finger to sock your eye (sockeye). The largest finger is the king or Chinook. Coho is the ring finger. Pink, the smallest salmon, is the pinkie. Now whenever you want to remember the names of the five Pacific salmon, all you have to do is look at your hand!

Chum and pink salmon are not found in Idaho. Wild coho are extinct in Idaho. Credit: Idaho Department of Fish & Game

The Journey of a Life Cycle!

> "The inland West's wild salmon awaken, at birth, to the pebbles and clear flow of a high mountain stream. The tiny fish thus bond not to a parent fish, but to the parenting stones and flow of their birthstream."
>
> ~ David James Duncan

Female Chinook salmon digging a redd with her tail in Johnson Creek. Credit Nez Perce Tribal Fisheries

Nesting & Hatching

Just like other animals who build nests for their offspring, wild Pacific Chinook salmon create a cluster of nests made out of gravel, called **redds**, as a place to lay their eggs in freshwater rivers and streams. From around August through December, each spawning-female digs a redd by laying on her side and flapping her tail, called her caudal fin, back and forth to move gravel on the **substrate** of the stream into a nest. Spawning-males instinctively find females who are ready to spawn and swim up beside them, coaxing them to release their eggs. When they are side by side, the female releases her eggs and the male covers them with a substance called milt, to fertilize them. This process may happen four or five times, with females depositing more eggs to be fertilized with milt into multiple nests. These dedicated female Chinook will protectively remain nearby in the river from around 4 to 25 days before they die, having fulfilled their lifelong destiny!

Chinook eggs incubating in our classroom. Credit: Deirdre Abrams

Remarkably, female Chinook in Idaho can lay up to about 5,000 eggs. It can take up to 12 weeks for eggs to hatch, depending on the water temperature. The colder the water, the longer they take to hatch. Although she lays thousands of eggs, I think the mother Chinook spawning in the wild in the Middle Fork of the Salmon River would be disappointed to know that only about 1% of her eggs will survive to become spawning adults. This is due to various problems, such as predation, low water levels, too much silt in the water, freezing, and, especially, going through the many perils of trying to get through all the hydroelectric dams and reservoirs.

Chinook eggs hatching into alevin in our classroom.
Credit: Deirdre Abrams

The wild eggs that do survive, hatch into **alevins**, or sometimes called sac-fry, by breaking free of their soft shell. They still live in their gravel nest for protection at this stage. Alevins' bodies, which are about an inch long, have an attached yolk sac, which provides a food reserve for their first month until the yolk sac is absorbed.

Freshwater Life

After alevins absorb their yolk sac in around four weeks, they emerge from their nests. At this stage, salmon are called **fry**, when they swim freely and begin feeding on plankton, nymphs, and larvae of insects. Salmon need a clean, flowing, aquatic habitat that has shade and lots of dissolved oxygen in order to survive in their early years. Trees play an important role in creating a healthy river habitat for Chinook while in freshwater. Standing trees create shade, keeping the water cool, and their roots stabilize the soil on stream and riverbanks to minimize erosion. Tree leaves make food for insects like mayflies, caddisflies, stoneflies, and other aquatic macroinvertebrates, which are **prey**, or food, for young salmon. When old trees topple into rivers and streams, they become the perfect, safe hiding place for young salmon. Deep pools also provide important hiding places for salmon to conceal themselves from predators like larger fish, snakes, bears, and birds. **Riffles** are areas of a stream or river where the water is fast moving, shallow, and has gravel or rocks at the bottom. Riffles offer the perfect

Development of alevin to fry. Credit: USDA Forest Service

set of conditions needed for salmon to lay their eggs and create an oxygen-rich habitat.

As salmon continue to grow to about four inches, they develop "parr" marks. These dark, vertical bar-shaped markings function as camouflage for small salmon. At this stage, young salmon are known as **parr** and begin adding larger aquatic insects to their diet.

Chinook: from bottom: fry, parr, smolt, juvenile. Credit: FISHBIO

Most fish can either live in freshwater or saltwater, but not both for a long period of time. Salmon are remarkable in this way, as their bodies can change to adapt from a freshwater home to a saltwater environment, and back again to freshwater! After their first year, Chinook parr get ready to move from freshwater toward the ocean, although fall Chinook (Chinook who begin their return migration back to Idaho in the fall) spend only one to three months in freshwater before heading to the ocean. Scientists believe that Chinook use microscopic **magnetoreceptors**, like a magnetic compass, to navigate the Earth's magnetic field to find their way to the sea and around the ocean. Various migratory animals such as migrating birds and anadromous salmon have a mineral called "magnetite" in the tissue within their skulls. Because "magnetite" is magnetic, salmon can use this as a way to navigate to the sea and back, also with the help of their ability to detect and remember differences in water chemistry in the river.

At this migration stage of their life cycle, salmon are called **smolts**, the teenagers of the salmon world, and are about as long as your hand. As smolts, their bodies prepare to make complex changes to survive in seawater and begin to add other fish and tiny shrimp to their diet. On the two-page map at the beginning of the book, use your finger to trace the freshwater from the Stanley Basin, the farthest reaches on the Salmon River in Idaho, to the Columbia Estuary on the Pacific Coast. This is the incredible journey that Chinook smolts, or juveniles, make when leaving their freshwater home from the farthest reaches in Central Idaho.

Chinook parr with the distinct vertical parr marks. Credit: Idaho Department of Fish and Game

Make Your Own Floating Magnetoreceptor...

You will need:
- a sewing needle or paperclip
- a round slice of cork or coin size piece of foam
- a magnet
- a shallow bowl or clear 8 oz. plastic cup with wide mouth
- water
- a real compass to check which end of your needle or paperclip is pointing north

1. Rub your sewing needle or paperclip with the magnet to magnetize it. If you want to test the needle's magnetism, see if it will attract another needle or pin. If the magnetized needle picks up the pin, it should be ready.

2. Carefully insert the magnetized needle through the slice of cork or piece of foam horizontally (or tape it on top) so that the needle will be parallel to the water. This is your "magnetoreceptor".

3. Fill a bowl or the wide plastic cup with a few inches of water and place the compass on the water. The magnetized needle will align itself with the earth's magnetic field to point north to south.

4. How do you know which end of the needle is pointing north? Use a real compass to check and use a marker to write "North" and an arrow pointing to it on the piece of cork or whatever you used as the floater. To use your magnetoreceptor again, instead of the real compass, just follow the arrow you drew to North!

During this migration, young salmon ride the spring run-off from the mountain waters of Idaho to the Columbia Estuary. They ride the river current backwards, facing upstream, as their birthplace fades to new sights and smells. During this outward migration period, salmon's bodies go through a physiological process called **smolt-ification**, which ends by the time they leave the estuary, headed out to sea. This process helps them prepare for the effects that saltwater will have on their bodies. Smoltification does not begin at a specific time in all salmon, but varies depending on temperature, size, rate of growth, age, or a combination of these factors.

The word **estuary**, a coastal tidal marsh connecting rivers to the sea, is derived from the Latin root "aestus", which means "tide" and is where juvenile salmon smolts live temporarily while their bodies get ready for life at sea. It is a nutrient rich environment where the coming and going of tides, twice a day, carries saltwater from the ocean inland to mix with freshwater rivers. Estuaries are a critical transition zone for salmon, where their bodies that previously resided in freshwater must adapt to living in saltwater. At the surface of an estuary, the freshwater is less dense than saltwater, so it floats on top. This is where salmon can spend time in familiar freshwater yet begin to acclimate to the saltwater beneath them.

Columbia River Estuary. Credit: US Geologic Survey (usgs.gov)

During their time spent in the estuary, salmon's bodies grow several centimeters due to the abundant nutrients there, their skin color changes, they become more buoyant and streamlined, and their bodies become able to regulate and balance salt.

Unfortunately, many smolts will not complete the journey to the Pacific Ocean, as they will become food for other animals. Today, Idaho-born Chinook face other challenges in their migration besides predators, which previous generations of salmon did not experience even 50 years ago. Today, these fish must navigate their way over or through the eight dams on the Snake and Columbia Rivers. Warm reservoirs at the dams provide no current, are disorienting, and are filled with larger, awaiting predator fish. Imagine riding the current backwards, just floating along, and then hitting stagnant water at each dam. You would need to turn around and figure out where to go each time, which is not in your genetics if you're a salmon. In fact, depending on the year, 7-15 percent of Chinook smolts die as they attempt to pass large reservoirs and cross over each dam.

Make an Estuary Water Model

You'll need: 2 clear plastic cups, warm water, 3 tablespoons of salt, a measuring tablespoon, blue & yellow food coloring.

1. Fill one of the cups halfway with warm water. Add blue food coloring and stir until the water turns to a medium blue color to represent the ocean. Now add 3 tablespoons of salt and stir until completely dissolved.

2. Fill the other cup halfway with warm water. Add yellow food coloring and stir. This will represent freshwater, so you will NOT add salt.

3. Using a measuring tablespoon, pour the yellow "freshwater" into the blue "ocean" water by placing the spoonfuls of freshwater against the inside of the cup so that it drips down the side slowly. Tilt the cup of "ocean" water as you slowly drip in the freshwater. Keep slowly adding the yellow freshwater to the saltwater until you see a separation of the two.

What do you notice when you look through the cup? Which is less dense (will float on top)-the saltwater or the freshwater? This area of mixed saltwater and freshwater is where salmon transition from living in rivers to living in the ocean!

Pacific Ocean Life

When juveniles have fully transitioned to live in saltwater, they move out of the estuary to the wide Pacific Ocean for most of their lives in order to feast on fish like herring and anchovies and some crustaceans. This abundance of food is needed to fatten them up for their long, eventual journey back to their birthplace to spawn. In the ocean, there are fewer places to hide, with no overhanging trees or logs like in freshwater streams. As a result, the color of their bodies changes during smoltification from a darker color with black parr marks to a silvery color to help them hide in the light conditions of the ocean's surface water.

Fishing boats on the Pacific Ocean. Credit: FISHBIO

As they continue to grow, salmon will live together in schools close to the shore for protection from predators such as orcas, harbor seals, and sea lions. As adults, Chinook will venture out further up the coast to the Gulf of Alaska, where they will face both predators and commercial fishing nets. Some Idaho Chinook swim up to 10,000 miles during their time at sea. If they survive this, they will remain in the sea until their instinct tells them that it is time to return to freshwater to spawn.

Return of the Kings

Adult Chinook salmon who have spent a sufficient time at sea develop an instinctive urge to spawn and know that it is time to make the journey back home. Biologists think that genetic impulse, day length, ocean conditions, and level of maturity trigger this urge to migrate home.

When salmon migrate back to freshwater to their birthplace to spawn, they must spend time in an estuary again to transform from saltwater organisms back to freshwater organisms.

As large groups of Chinook salmon gather in the estuary waters to migrate together, they begin to darken in color. Male salmon will develop a hook-like jaw, called a kype, and sharp canine-like "breeding teeth" to help them compete with other males during spawning.

Making the journey back to their spawning grounds is not an easy task. Chinook stop eating after they begin their migration home, and their stomachs begin to deteriorate since they are no longer needed. This makes more room for maturing reproductive organs in both females and males and to host thousands of eggs in the females. All the weight they gained in the ocean pays off, as Idaho-born Chinook will live off the fat on their bodies for energy as they begin their upstream journey.

During their return to the streams where they were born, adult Chinook must jump waterfalls, some 12 feet high, swim up fish ladders over eight dams, and navigate warm,

Chinook salmon migrating home to spawn. Credit: FISHBIO

Credit: Bonneville Power Administration (bpa.gov)

still reservoirs. To make things worse, they must dodge predators, like fisherman, bears, terns, seals, and sea lions as they migrate. One notable problem in their return migration is the lurking sea lions who swim up the Columbia River from the ocean to the Bonneville Dam. This is where they prey on large numbers of salmon who are crowded at the bottom of the fish ladders. It's no wonder why so few survive this return migration.

How do Chinook find their way back to the river or stream section where they were born? Adults returning to their spawning grounds rely on chemical cues of the water as they sniff their way, so to speak, to find the right area. They build their 'smell memory-bank' when they start migrating to the ocean as young fish, and this guides them back home, like a breadcrumb trail, to the body of water where their lives began. In fact, Chinook have been detected wandering into other tributaries yet turn around when they realize that stream doesn't have the water-chemistry smell that they remember. A few salmon, however, do stray sometimes as a way to seed new habitat. When Idaho Chinook travel from the ocean to return to their river-birthplace to reproduce, they often will come within yards of where they were born. Salmon know that the stream where they were born is a perfect breeding area because, well- they were born there- and are still around! As a result, wild salmon won't spend much time looking for a different area to spawn unless those spawning grounds become inaccessible for some reason.

Salmon swimming up a fish ladder at a dam. Credit: FISHBIO

Salmon do not eat during the migration back to their birthplace to spawn. They rely on their fat stores from eating other fish and crustaceans in the ocean to make the long journey home.

Chinook who return to spawn in Idaho arrive at different times. "Spring Chinook" depart the ocean in spring and arrive in Idaho the earliest. They spawn in the upper and intermediate reaches of the Salmon and Clearwater Rivers in late August through September. "Summer Chinook" leave the ocean in early summer and spawn in the intermediate river stretches like the South Fork of the Salmon River during the same timeframe as spring Chinook. The South Fork of the Salmon River is exclusive to summer Chinook spawning. "Fall Chinook" head to Idaho in late summer and are the latest spawners. They reproduce around September through December. Fall Chinook are also the largest Chinook to return to Idaho, which is why they are capable of spawning in the big, deep rivers of the main stem of the Snake River and the Clearwater River.

When those Chinook who actually survive the treacherous journey to their spawning grounds make it to their birthplace, their bodies have drastically changed. Their color shifts from the silvery color that they wore at sea to a range of a yellowish-olive, maroon, to greyish-black color. Another change in their appearance when they arrive at their spawning grounds is their size. Because they stop eating when they leave the ocean, both male and female Chinook who return the great distance to Idaho lose around 20% or more of their body mass. In contrast, some appearances that remain the same in both freshwater and ocean habitats are that Chinook always have black spots on the upper part of their body and tail and have completely black mouths, unlike coho salmon, who have completely white mouths.

Salmon Run	Pacific Ocean to Idaho Migration	Spawning in Idaho	Idaho Spawning Grounds
Spring Chinook	Leave ocean in spring	Around late August or early September	Upper and intermediate reaches of Salmon & Clearwater Rivers
Summer Chinook	Leave ocean in early summer	Around late August or early September	Intermediate reaches of Idaho rivers-South Fork of Salmon River is exclusive to summer Chinook
Fall Chinook	Leave ocean in late summer	Around September-December	Main stem of Snake, Salmon, & Clearwater Rivers

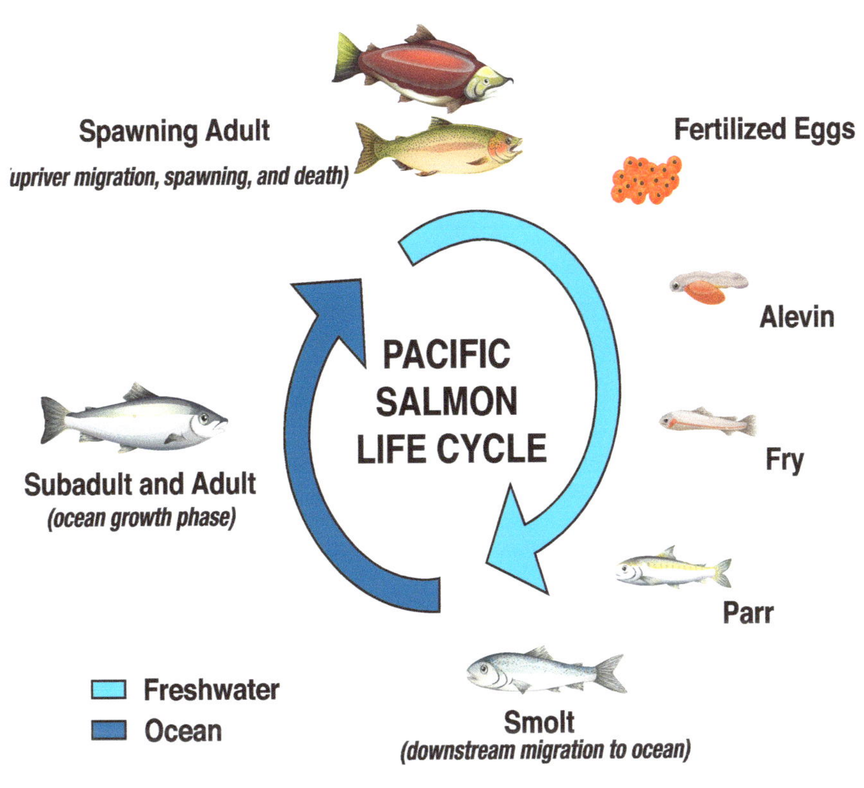

Spawning Adult
(upriver migration, spawning, and death)
Fertilized Eggs
Alevin
PACIFIC SALMON LIFE CYCLE
Fry
Subadult and Adult
(ocean growth phase)
Parr
Freshwater
Ocean
Smolt
(downstream migration to ocean)

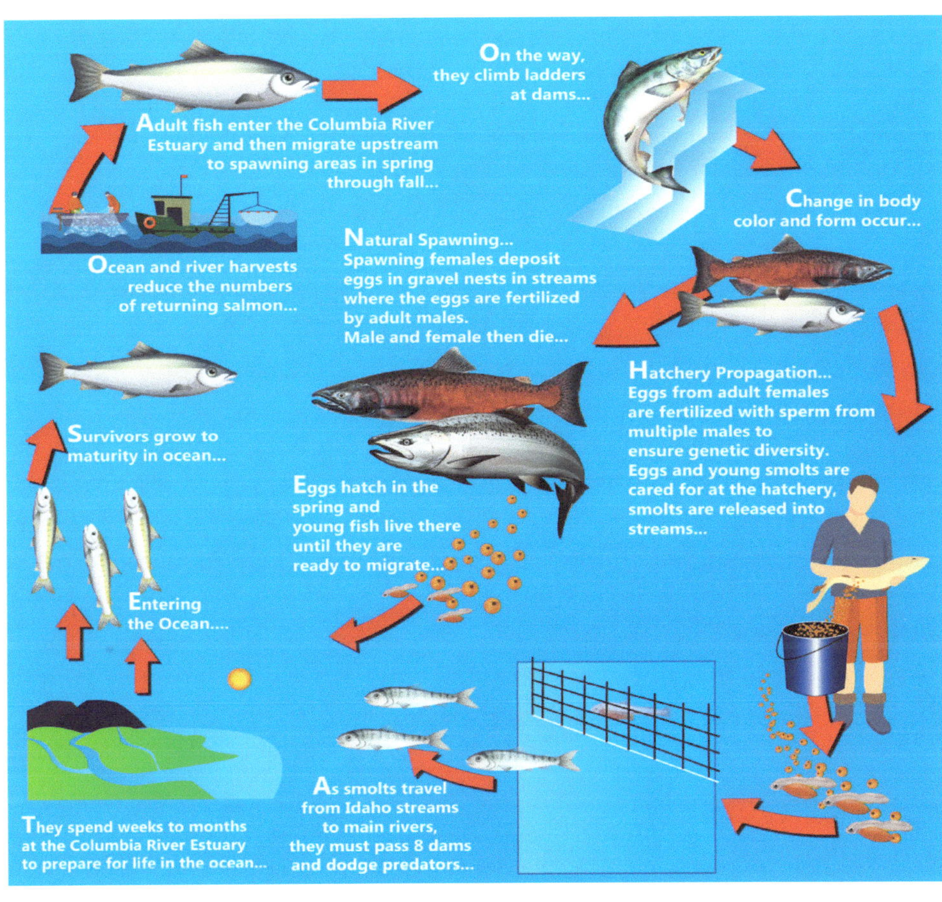

On the way, they climb ladders at dams...
Adult fish enter the Columbia River Estuary and then migrate upstream to spawning areas in spring through fall...
Change in body color and form occur...
Ocean and river harvests reduce the numbers of returning salmon...
Natural Spawning... Spawning females deposit eggs in gravel nests in streams where the eggs are fertilized by adult males. Male and female then die...
Hatchery Propagation... Eggs from adult females are fertilized with sperm from multiple males to ensure genetic diversity. Eggs and young smolts are cared for at the hatchery, smolts are released into streams...
Survivors grow to maturity in ocean...
Eggs hatch in the spring and young fish live there until they are ready to migrate...
Entering the Ocean....
They spend weeks to months at the Columbia River Estuary to prepare for life in the ocean...
As smolts travel from Idaho streams to main rivers, they must pass 8 dams and dodge predators...

Spawning: Tying the Sea to the Mountains

Male and female Chinook salmon spawning. Credit: FISHBIO

The reason Chinook and other salmon migrate back to where they were born in Idaho is to fulfill their part in the continuation and survival of their species, just like all living organisms must do. This stage of life is when Chinook salmon come full circle in their life cycle, ready to create new life and end their own.

Like you read earlier, females dig their redds with their tails while waiting for the right male to come along. Males with their distinctive kypes, or hooked jaws, will compete with one another to breed with the female, with the largest males often winning-over the female. A female Chinook will dig several nests in which to lay her eggs. When she has deposited all of her eggs and they have been fertilized, she will guard her offspring for days while her body slowly disintegrates. Here, the male and female Chinook salmons' lives will often end in the same stream where their life began. When their offspring hatch, the newly born Chinook will be born in the same waters they, too, will die in if they make it home. It is both an end and a beginning for these amazing fish.

Chinook salmon carcasses provide marine nutrients for Idaho streams and land. Credit: FISHBIO

This death and decay at the end of the Chinook salmon life cycle in Idaho literally ties the ocean to the mountains! The marine nutrients that were consumed by these special salmon in the ocean become nourishment in the streams of Idaho where they die. Their carcasses provide rich nutrients for other animals and also nourish the roots of trees and plants on streambanks. Even the Chinook who die elsewhere, and do not make it back to spawn in Idaho, provide nutrients to animals and plants in the place where they die. These marine nutrients would not exist as part of the food web in Idaho if it weren't for the migration of salmon.

Web of a Special Life

> "Abundant salmon returns feed the rivers and shape the habitats that support the next generation of wild fish." ~Guido Rahr

In the food chain, Chinook salmon, classified as **consumers**, feed on plankton when they are in the fry stage. Plankton are classified as producers. **Producers**, such as plants, make their own food through the process of **photosynthesis**, whereas consumers must eat producers or other consumers for food. When young Chinook reach the parr stage, they begin to eat insects. Insects are also consumers because they eat plants or other insects. As adults living in the ocean, salmon will eat crustacean larvae and smaller fish, which are also both consumers. Check out the web of life diagram on the next page. You can see that Idaho Snake River Chinook salmon are an important food source for Southern Resident killer whales of Puget Sound (SRKW), also called orcas, who reside off the coast of Oregon and Washington. At least 50 to 80% of these orcas' diet consists of Snake River Chinook salmon in the winter and spring. The average adult Southern Resident orca must consume almost 30 adult salmon daily in order to just get the energy needed to survive. This makes Chinook salmon a **keystone species**. A keystone species is an organism upon which other species in an ecosystem are dependent for survival. In other words, if that keystone species were removed, it would have drastic, negative effects on the entire food chain in the ecosystem.

After they spawn and die, both male and female Chinook salmon become important food for other consumers in Idaho. Bears, wolves, eagles, ospreys, and other animals feast on the carcasses of dead salmon in the streams where they die. Those animals then disperse each salmon's nutrients into forests and meadows by dragging them into the woods to eat or through the digestive process. Because Chinook salmon are the largest salmon, they provide the largest amount of biomass (organic matter) per salmon in Idaho, creating a huge network of energy and nutrients throughout the ecosystem. Did you know that eating one serving of Chinook salmon gives you 20 grams of healthy fat, 30 grams of protein, and a huge number of vitamins and minerals? That same serving also provides over 33 grams of omega 3 fatty acids, which help to lower the risk of heart disease, depression, arthritis, and dementia. This nutritional treasure is one of the reasons salmon were, and still are, vital to indigenous peoples' culture and important to all of us for our health!

On a micro-scale, **decomposers**, like bacteria and maggots (fly larvae), break down dead Chinooks' flesh, enriching the waterbody, which then provides marine nutrients for Chinook's offspring when they hatch and for the roots of trees and other plants that line the stream wherever it flows.

Notice the extensive transfer of energy as it moves between ecosystems, beginning with the sun.

PACIFIC CHINOOK SALMON WEB OF LIFE

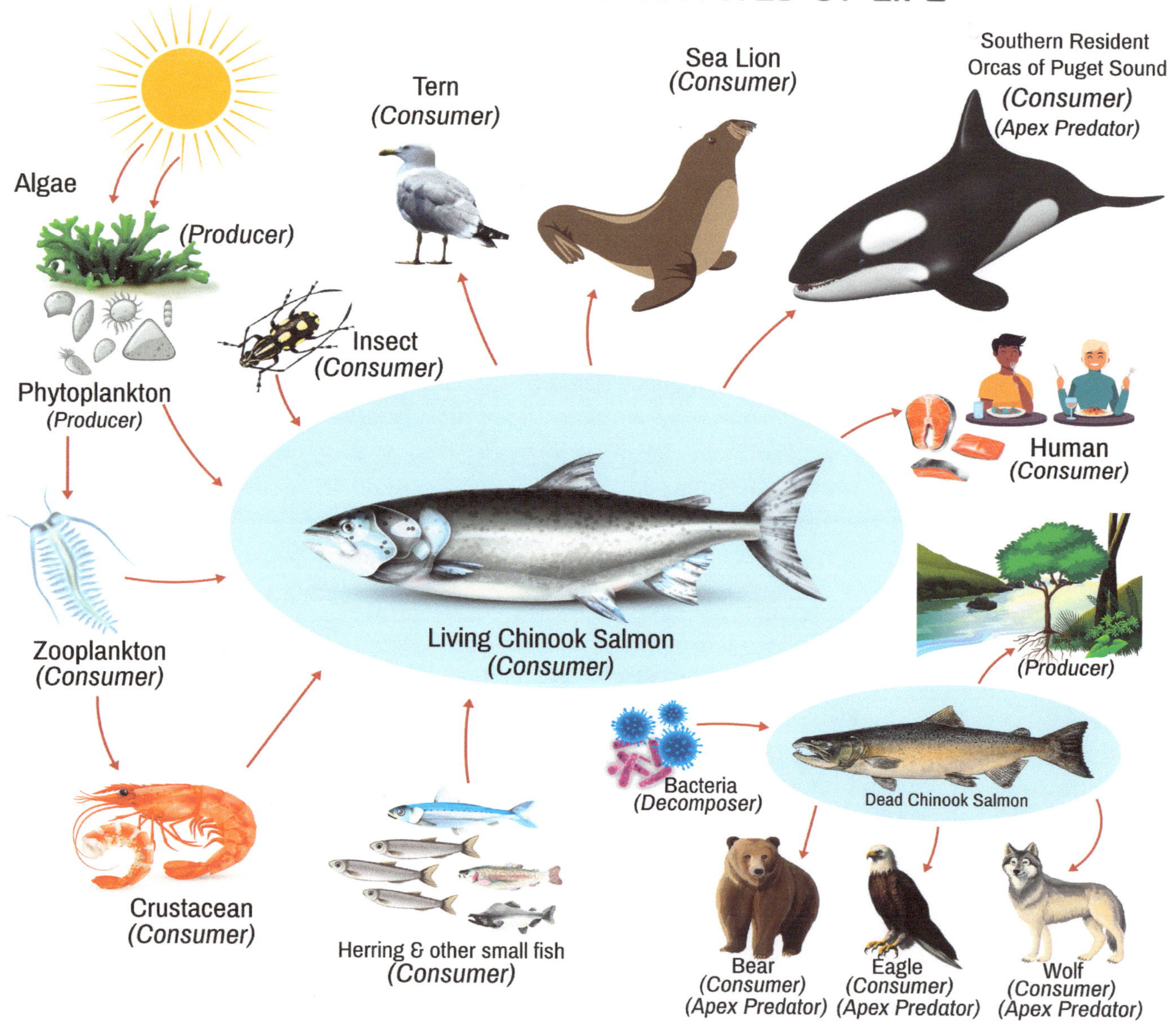

Anatomical Structure of a Chinook Salmon

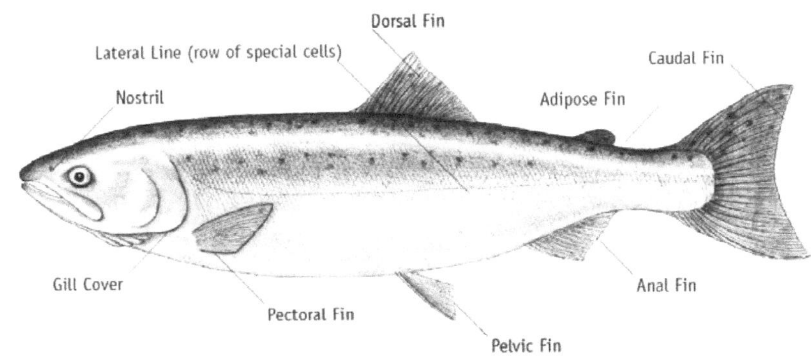

Credit: Alaska Department of Fish & Game

Structure & Functions of External Anatomy

Eyes

Salmon depend on eyesight to see food, avoid predators, and to navigate. Fish do not have eyelids because their eyes are always bathed in water and do not need tears.

Nares

Chinook have small holes, called nares, that look like nostrils on their head. These nares support their strong sense of smell, which they use to remember the distinct water chemistry of the stream in which they were born. When returning to that natal stream to spawn, they use this **"olfactory"** sense of smell to find it, even through up to 900 miles of river! Smell can also be used to find food and avoid predators. Salmon do not use their nares to breathe like other animals do. They use their mouths and gills.

Lateral Line

Salmon do not "hear" like other animals. Low frequency sounds are detected in the water through the lateral line, a system of fluid-filled sacks with hair-like sensory features that are open to the water through pores along the side of the fish. Salmon do not have ears, so the lateral line allows them to detect movement of other fish and predators in the water.

Mouth

Salmon use their mouths to catch and hold food, but do not chew before swallowing. They swallow food whole. The function of the teeth in their mouths is to keep their prey from escaping. You can tell what kind of food a fish eats by looking at its mouth and teeth. For example, salmon have backward-pointing teeth on their gums. Salmon also have teeth on their tongues and the roof of their mouths. These teeth form a one-way path for wiggly, slippery food that is swallowed whole.

The jaws of male Chinook are much more pronounced than females. The males' kype (jaw) becomes much more hooked when they are in their spawning stage. Because their bodies are already breaking down during spawning, their gums recede, making their teeth look larger and scarier to other males.

Notice the hooked "kype" (jaw) of the male Chinook.
Credit: Roger Phillips, Idaho Department of Fish and Game

Opercula (Gill Covers)

Salmon gills are covered by gill covers, called opercula, which protect the gills. Each operculum (one on each side) forces water containing oxygen over the gills, with the help of the mouth, so fish can "breathe".

Fins

Chinook salmon have two sets of **paired fins**. The pectoral fins are located at the front of the body and the pelvic fins are located toward the back of the body, one on each side. These fins are basically used to maneuver and balance. Salmon also have four single fins. The **dorsal fin** is the large fin located on top of the fish. The **anal fin** is located underneath the tail. These fins are also used for maneuvering and balancing. The **caudal fin** is also known as the "tail" and is the most important fin. It is used to propel the salmon and also acts as a rudder. Female salmon use this fin to dig the redd where she deposits her eggs. Lastly, the **adipose fin** is located on top of the fish between the dorsal fin and the tail. Hatcheries often remove this fin so that they can tell if it is a hatchery fish, not a wild fish. It

School of Chinook. Credit US Fish & Wildlife (usfw.gov)

was thought that this fin served no purpose, but according to The Fisheries Blog, "Research suggests that the adipose fin may serve as a 'precaudal flow sensor' to improve maneuverability in turbulent waters. This new research may raise concerns because adipose fin clipping is commonly used to "mark" hatchery reared salmon to distinguish them from wild salmon for catch and release management purposes."

Scales

The scales' function on a salmon are to protect their bodies like a coat of armor. They grow in circular patterns and can be used to determine the age and life history of a fish. Under a microscope, scales show growth rings which, like the growth rings of a tree trunk, can be used to estimate age. Fish don't grow more scales as they get older; the scales just get bigger. The scales, which are formed in the skin and overlap like shingles on a roof, protect against abrasion, disease, and predators. A layer of mucous, or slime, covers the scales which further protects fish from disease, fungi, and viruses, and also helps fish to glide through water more efficiently and escape from predators.

Scales can identify the age of a salmon, much like tree rings identify the age of the tree.

Structure & Functions of Internal Anatomy

Ovaries

Female salmon have a reproductive organ called ovaries which produce eggs. Once fertilized by a male salmon, these eggs will grow to be salmon offspring. If a female produces four thousand eggs, only about two will make it back to their birthplace to spawn as adults.

Testes

Male salmon have a reproductive organ called testes, which produces milt. Milt is deposited on salmon eggs to fertilize them.

Liver

Like humans, the liver stores, synthesizes, and secretes essential nutrients from food. It destroys old blood cells and maintains proper levels of blood chemicals and sugars. The liver also secretes enzymes that break down fats to assist digestion.

Gall Bladder

This organ is a sac attached to the liver in which bile is stored and used to digest fats.

Heart

Salmon have a two-chambered heart, unlike humans, who have a four-chambered heart. This organ circulates blood throughout a salmon's body and is located between its gills.

Esophagus

Sometimes called the gullet, the esophagus carries food from the mouth to the stomach. Like us, salmon have a sac-like digestive organ which receives food from the esophagus. The stomach uses digestive juices to break down and deliver food to the intestine.

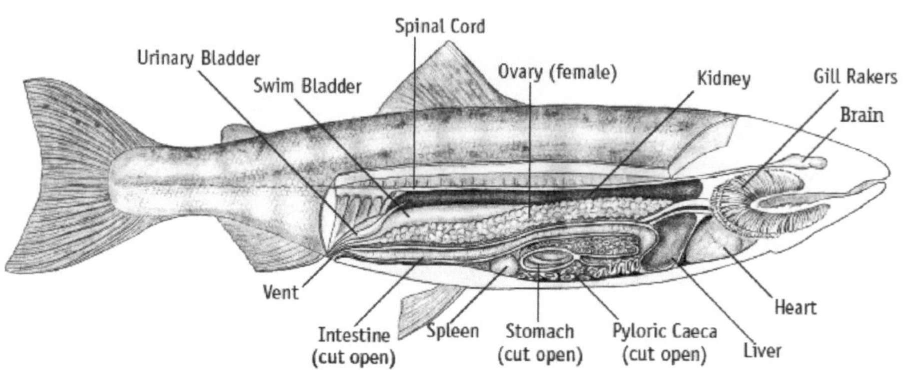

Internal anatomy of a female salmon. Credit: Alaska Department of Fish and Game

Pyloric Caeca

Pronounced *"pi-lor-ik see-kuh"*, salmon have an intestinal pouch of the digestive system which absorbs nutrients into the blood.

Intestine

The intestine of a salmon is a section of the digestive tract between the stomach and the anus where nutrients are absorbed and waste is transported to the anal vent.

Anal Vent

This is where urine, feces, eggs, and milt exit the digestive system.

Kidneys

In salmon, the two kidneys are joined together, unlike ours, which are separate. Kidneys are extremely important organs for anadromous fish, as they assist in "osmoregulation", which is the ability to control the concentration of substances, like salt, in body fluids compared to the water outside the fish. The kidneys, in this way, are key to their transition from freshwater to saltwater and back. Also, kidneys in a salmon remove waste from the blood and produce urine.

Male Chinook Salmon, Idaho, 2017. Credit: Roger Phillips, Idaho Department of Fish & Game

Spleen

A salmon's spleen stores blood for emergencies and recycles old red blood cells. Another purpose of the spleen is to produce white blood cells to help protect the salmon against disease and infection.

Gills

Fish have a special way to get oxygen out of the water they live in – gills. Gills are covered by a gill cover, called the operculum, which protects the gills and also forces water, containing oxygen, over the gills with the help of the mouth. The fish takes in water through its mouth. The dissolved oxygen in the water is absorbed through the membranes in the gills into the blood vessels. Carbon dioxide is released into the water across the same membrane. Chinook gills are very delicate yet are far more efficient than human lungs because they can extract up to 80% of the oxygen dissolved in water. Human lungs only extract up to 25% of oxygen from the air with each breath.

Brain

This organ coordinates the messages received about the environment from the sensory organs , like eyes, lateral line, and nares. The forebrain controls smell, the midbrain controls vision and learning responses to stimuli, and the hindbrain coordinates movement, muscles, and balance of the salmon.

Otolith

Referred to as the ear bone, these calcium carbonate structures help keep fish balanced and upright in water by detecting gravitational pull. Each salmon has three sets of otoliths. The growth rings formed on these round "bones" can determine the age of the salmon, just like the scales.

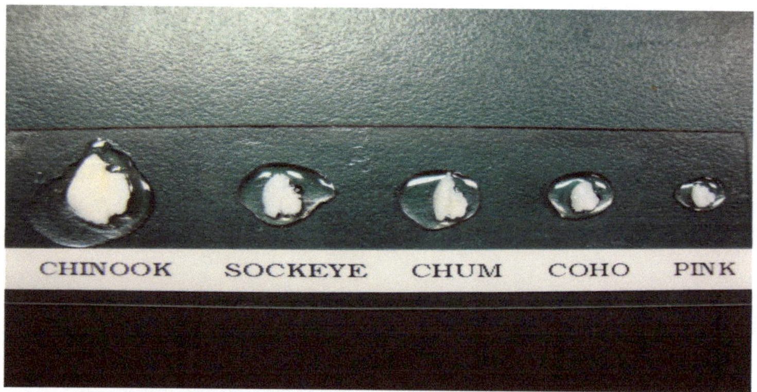

Otoliths from different sizes of salmon. Credit: Alaska Sustainable Salmon Fund

Importance to Native Culture

"*The salmon are one of our best teachers. We learn from them that we have to do certain things by the seasons. We watch the salmon as smolts going to the ocean and observe them returning home. We see the many obstacles that they have to overcome. We see them fulfill the circle of life, just as we must do. If the salmon aren't here, the circle becomes broken and we all suffer.*"

— Leroy Seth, Nez Perce Tribal Elder

Map of Nez Perce Historical Trail. Credit: National Park Service

Tens of thousands of tribal people once lived beside the mouth of the lower Columbia River, subsisting largely on salmon. These were the tribes of Chinook people, whose ancestors lived in this area for at least 9,000 years before explorers arrived in the 1700s. When Meriwether Lewis and William Clark traveled to the area in 1805, they met the Chinook people and ate salmon with them. Lewis and Clark wrote about a native marketplace by the river that centered on the trade of salmon, and they watched the Chinook festival to celebrate the first catch of the season, seeing how connected these **indigenous**, or native, people were to the salmon.

The *Nimiipuu* (NEE-me-poo) people, known today as the Nez Perce Tribe, have relied heavily on salmon and steelhead for spiritual, cultural, and nutritional purposes. The Nez Perce live in North Central Idaho today, yet their traditional territory spanned from Montana, Wyoming, and Idaho to Northeast Oregon and Southwestern Washington. Nez Perce families lived along streams and rivers of the Snake and Clearwater Rivers where salmon were found in plenty. They dwelled in longhouses during winter and **tepee** structures, which were more portable, when traveling to hunt, fish, or to trade with other tribes. Chinook and other salmon and steelhead were their main food source traditionally, besides wild game, like buffalo, blackberries and huckleberries, and roots. Today, Snake River salmon are still an important food source and cultural symbol to the Nez Perce.

Historically, in late May and early June, the rivers that the Nez Perce fished were filled with eels (lamprey), steelhead, and Chinook salmon. The Nez Perce were well known for their breeding of Appaloosa horses-which is Idaho's state horse today-and this helped them travel to other places to fish and trade. In the spring, they went to shared fishing sites along the rivers of the Columbia River Basin to trap, spear, or haul fish in with large dip nets. Traditionally, the men in the tribe fished and the women dried and stored the fish.

One method of harvesting salmon was the use of a spear, which had a wooden shaft with a fixed, sharp point at the end, made of bone or wood with three barbs that looked like giant forks. Nez Perce fishermen also used harpoons, which had three forks with a connected rope. The fisherman would quickly stab a passing fish, and the point would stay in the fish but would slide off the shaft. As the fish tried to swim away, it was stopped by the rope that still connected the harpoon tip to the long shaft. He could then pull the fish in, much like modern fishermen reel in a fishing line. The Nez Perce also built fish traps and fished for salmon from dugout canoes. When salmon returned in large numbers to spawn, there were so many that the fishermen could also simply dip a large net, made of plant fibers, into the river and wait for the salmon to swim into it.

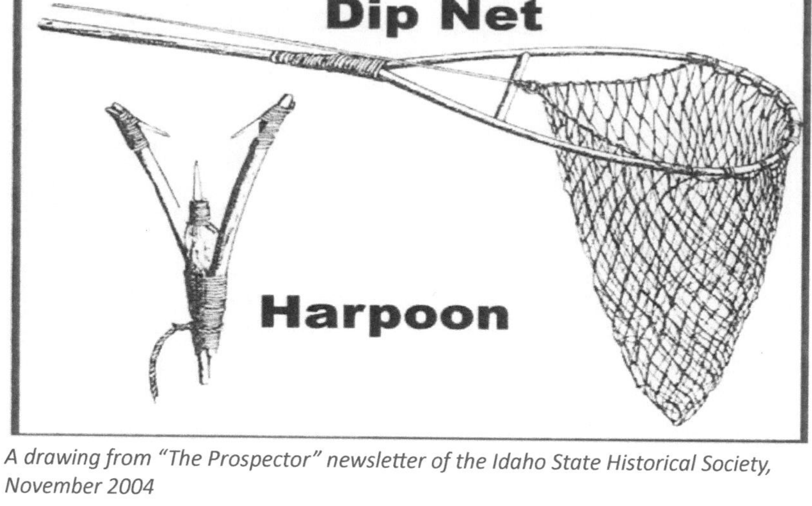

A drawing from "The Prospector" newsletter of the Idaho State Historical Society, November 2004

The first fishing of the season has always been marked by rituals and a ceremonial feast known as kooyit, or First Salmon Feast, which is still part of the Nez Perce tradition today. Gratitude is offered to the Creator and to the fish for returning and giving themselves to the people as food. Anthony Capetillo, a Nez Perce Tribal member and descendent of Chief Joseph's band, says, "Along with praying for healthy returns of fish in the future, we also pray for the other foods involved in the ceremony such as the roots, meats, and berries. But the most important of all is the water because water is life. Water is the first and last thing we consume when opening and closing our ceremonies."

In the 19th Century, after Lewis & Clark visited the Nez Perce, fur trappers and traders and missionaries began moving into the area. By the 1840s settlers began traveling

Credit: Library of Congress, Washington, D.C.

through the area on the Oregon Trail, causing the Nez Perce lifestyle to change drastically due to competition for resources and hostile actions from the settlers. To protect their people, part of the tribe agreed, reluctantly, to the 1855 **Treaty** of Walla Walla, which granted them a **reservation** on about 7.5 million acres of their historic land and the right to hunt and fish on about 5.5 million acres that they were pushed to give up to the federal government. This forced agreement went against everything that the Nez Perce stood for, as indigenous people did not have an understanding of "owning" property prior to the arrival of white explorers and settlers. They dwelled on land where they lived in harmony with nature, valuing the exchange of giving and receiving between humans and other living things without owning them.

To make matters worse, in 1860, when gold was discovered on the Salmon and Clearwater Rivers in Idaho, thousands of miners and other settlers moved into the area. A renegotiation of the treaty was forced upon the Nez Perce, which greatly reduced the size of their reservation. Later, settlers, called "squatters", would begin to **homestead** on Nez Perce land even though the treaty specifically indicated that no American, non-tribal member, was allowed on the reservation without consent of tribal leaders. Several bands of the Nez Perce, given the name "Non-Treaty Indians" refused to give up their familial lands because it was a violation of the Treaty of Walla Walla. They would rather fight than leave this sacred ground where their ancestors were buried. This eventually led to the Nez Perce War in 1877. They were praised for their skilled fighting abilities, bravery, integrity, and for winning many battles, yet numerous Nez Perce men, women, and children lost their lives in this war.

After so much suffering and loss within his tribe, Chief Joseph ultimately proclaimed that his people would fight no more when he was promised that they would be allowed to return to their land. On that solemn day, he proclaimed, "Hear me, my chiefs! I am tired. My heart is sick and sad. From where the sun now stands, I will fight no more forever." After the war, the commanding

Nez Perce man in 1905. Credit: Library of Congress, Washington, D.C.

Celilo Falls on the Columbia River in 1931, where many indigenous tribes gathered to trade and celebrate salmon, before it was flooded to become the Dalles Dam. Credit: Library of Congress, Washington, D.C.

general of the U.S. Army overruled the agreement that had been made with the Nez Perce, and they were eventually moved to a reservation in Kansas, and then to Oklahoma as prisoners of war, rather than being returned to their land as agreed. After many broken promises and many lives lost, in 1884, Nez Perce families were finally able to go home to the Pacific Northwest where they could be united with their homeland, culture, and Snake River salmon runs, although it would never be the same. Some Nez Perce moved to the Lapwai Reservation in Idaho, and others, including Chief Joseph, moved to the Colville Reservation of confederated tribes in Washington.

The building of dams also drastically changed the lives of the Nez Perce and other tribal cultures. In 1957, the building of the Dalles Dam at Celilo Falls on the Columbia River destroyed Celilo permanently. This is where the Nez Perce and other tribes would gather each year to fish, trade, and socialize, and was a special part of tribal culture. Over time, the rest of the eight dams that were built, which harm migrating fish, has further devastated their culture and ability to harvest salmon.

In the early 1980s, The Nez Perce Tribe established a fisheries department, in which their mission has been to protect and restore salmon and steelhead according to their belief system about the natural world. They co-manage fish hatcheries with U.S. Fish & Wildlife Service (USFWS) and Idaho Department of Fish and Game (IDFG), but the Nez Perce Tribal Fisheries (NPTF) brings a unique perspective to managing fish as a cultural resource. The federal focus (USFWS) is mainly on preventing extinction. IDFG's focus is to "preserve, protect, and perpetuate" fish and wildlife. But at the same time, the agency is funded by hunters and anglers, so IDFG manages fisheries to provide sport-fishing harvest as their main goal when possible. In contrast, the Nez Perce Tribal Fisheries' goal is to fully restore salmon and other fish runs to their natural and historical numbers.

The Nez Perce, like other indigenous people, have many stories that have been handed down over time, which were created to convey the tribe's history, explain spiritual beliefs or lessons, or make sense of how things came to be. Many stories told involve animals like coyote, who is often the teacher or trickster, and salmon, which have fed their people throughout history.

Nez Perce Legend

In ancient times, Coyote came to the region that we now know as the Pacific Northwest. He saw that humans were in need of some help, so he created the mighty Columbia River. He connected this river hundreds of miles inland from the Pacific Ocean to a pond where women had been keeping two fish. Coyote declared that these fish would be "the people's food" because he knew the river would bring more humans. As the mighty Columbia flowed to the sea, the fish began to swim to the ocean and back again as salmon. Coyote then taught the people how to catch and cook these animals to feed themselves and to always give gratitude to the salmon for giving their lives to the people.

-adapted from Animals in Sea History by Richard King

King-Sized Problems & Critical Solutions

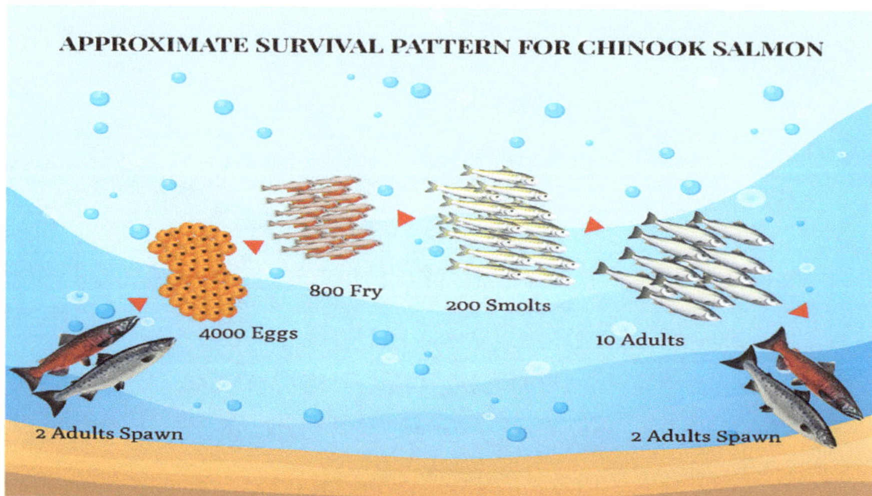

Approximate survival cycle of Pacific Salmon. 2020

Human impact on ecosystems has costs. The future has not been looking bright for these legendary fish who used to return to the high mountain rivers of Idaho in huge numbers. Now, before they complete their migration to the sea, Chinook must face many human-caused obstacles.

Due to the decline in both spring/summer and fall-run Chinook populations in Idaho, they were listed as "threatened" under the Endangered Species Act (ESA) in the 1990s, along with Snake River steelhead and sockeye. After 100 years of overfishing in the past, habitat destruction, dam building, water pollution, and warming acidic waters due to Climate Change, only an estimated 1,700 wild spring/summer Chinook returned to Idaho in 2020. This is a far cry from the millions who once navigated their way home.

The most widely accepted estimate is that 10 million to 16 million adult salmon and steelhead returned annually to the Columbia Basin prior to the 1930s. In the 1950s, 45,000-50,000 wild spring-summer Chinook spawned in the Salmon River in Idaho. For the past ten years, only about 16,000 wild spring/summer Chinook have returned to Idaho, on average, and only an average of 55,000 Hatchery spring/summer Chinook have returned in the last decade. "This year's adult returns to both the Clearwater Basin and Rapid River

Salmon obstacles. Credit: NOAA Fisheries

Hatchery are the lowest since we have been collecting PIT tag data," said Joe DuPont, Fisheries Regional Manager for Idaho Department of Fish & Game (IDFG) in June, 2020. IDFG biologists are increasingly concerned that they won't have enough salmon returning each year to supply **broodstock**, or reproducing adults, for the hatcheries that account for around 80 percent of the run. The Chinook who are born in the wild make up only about 20 percent of the run, and they are facing extinction.

Dams

A **dam** is a barrier constructed to hold back water, forming a **reservoir** that forces water through turbines to generate electricity. Beginning in the 1930s, **hydroelectric dam** power began to take root in the Columbia River Basin as a less expensive and less pollution-causing alternative to other resources in energy production. Dams were, and still are, celebrated by some as an innovative way to harness the power of rushing water, help farmers with the use of irrigation from controlling the flow of water from the reservoirs behind the dams, and to provide transportation of resources on cargo ships. Dam engineers designed **locks**, or ship passageways, which allow cargo ships to bypass dams to carry important resources, such as grains, from Idaho to the coast.

The problem is that dams are enormous obstacles that often kill or injure migrating fish. Salmon evolved to migrate in a free-flowing river. The conditions they face now are very unnatural to them. When these dams were built, biologists really only thought about the impacts to adult salmon and underestimated how lethal the dams would be for juvenile Chinook migrating downstream to the ocean. Another problem is that when dams were being built, some people thought that the need for alternative forms of energy, irrigation for crops, river transportation, and the jobs it would create were more important than the ecosystem that salmon belong to and critical habitats it would damage.

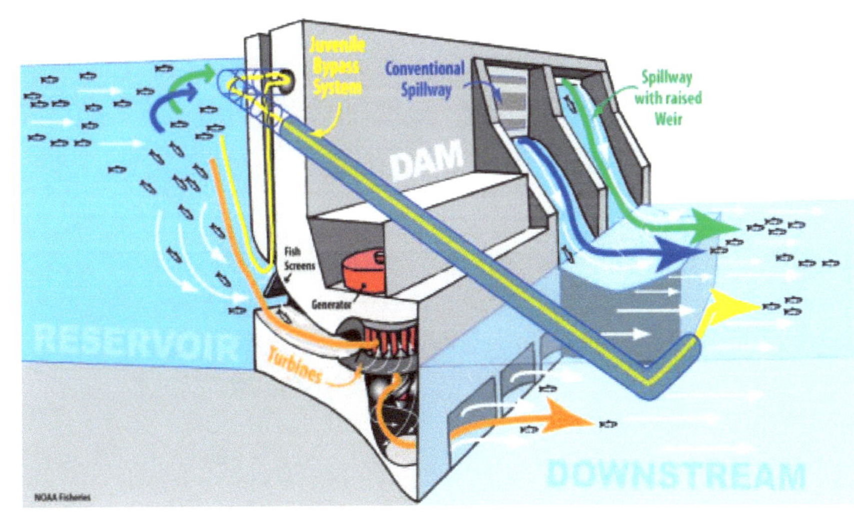

Dam Bypass Systems. Credit: NOAA

Today, Idaho's Chinook salmon have a far more deadly migration than the days before the dams blocked passage. The vast, warm reservoirs above the dams lack a current, causing juvenile salmon to get disoriented and become easy prey to predators, like Smallmouth bass and Pikeminnow. Prior to the construction of the dams on the Snake River, Idaho Chinook took a ride on the swift spring currents to the ocean. The rivers, at that time, moved from six to ten miles per hour, transporting Chinook to the sea in about a week or so. These days, due to over 300 miles of reservoirs, Chinook must swim much of the journey instead of being carried on the current. It now can take them months to reach the Pacific Ocean from Idaho instead of days. This creates many challenges for fish whose bodies are preparing to live in saltwater but have been delayed by these dams and are still living in freshwater. Smoltification is a stressful process and can begin to reverse if their migration is delayed for too long. This can be fatal for salmon.

Those smolts who do make it through the reservoirs to the dams must deal with absorbing too much nitrogen-filled water after plummeting over the dam spillway, navigating around the blades of the hydroelectric turbines, or getting routed into a tank to be transported by truck to the other side of the dam. Some Chinook die directly from these hazards, but many biologists believe that a large number of them who actually make it to the estuary experience "delayed mortality". This means that they die later from the collective stresses and possible delay in smoltification, caused by confronting all eight of these dams. According to Allison Guy, from Oceana, "Depending on environmental conditions, these dams may claim 50 percent of Chinook smolts as they migrate to the sea."

Again, these dams don't just cause problems during the outward migration. As adult Chinook make their way back to Idaho to spawn, they must navigate the same dams, falling prey to predators as they become concentrated while waiting to climb up fish ladders. Getting around the dams is a huge physical challenge for salmon. And again, they must find their way through slow-moving, predator-filled reservoirs upward of the dams.

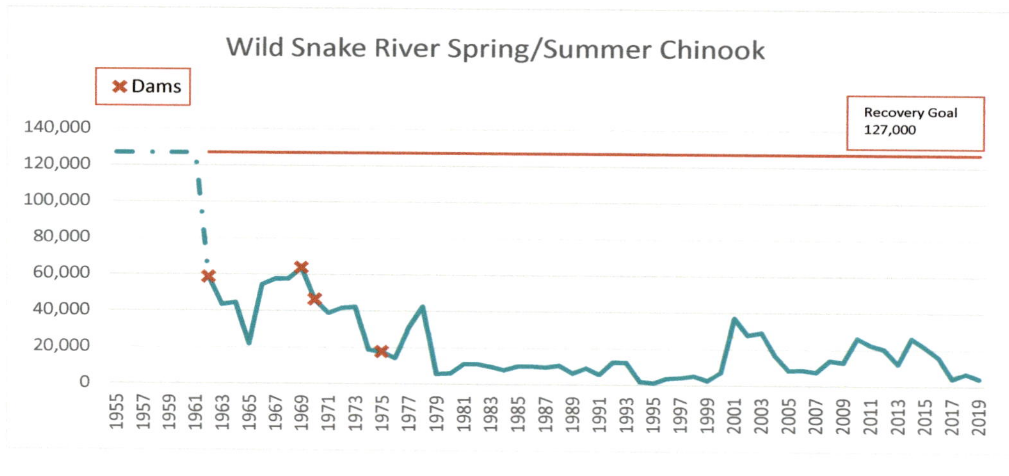

Credit: Idaho Rivers United

Many parts of Idaho have the most unspoiled, prime habitat for salmon in the world, thanks to many miles of Idaho wilderness and roadless areas with cold, high-mountain freshwater, yet these fish face too many obstacles getting home to spawn in this pristine habitat. Justin Hayes of Idaho Conservation League gives a flawless analogy: "It's like having a perfectly made up five-star hotel on the top of a mountain. The mint is waiting on the pillow, but 100 miles away someone is turning customers away on the road."

Salmon & Climate Change

McNary Dam, with fish ladder on left, one of the four Snake River dams causing problems for Idaho's Chinook salmon migration. Credit: salmonrecovery.gov

Climate Change, also known as Global Warming, is thought to be another impact from human activities and is affecting everything from the smallest of insects to the melting of the largest polar ice caps. Simply defined, Climate Change is the result of burning too many fossil fuels, such as oil, gas, and coal, which release unhealthy levels of carbon dioxide (CO_2) into the Earth's atmosphere. June of 1988 marked the date on which Climate Change became a national issue in the United States. NASA claims that nineteen of the twenty warmest years for the planet have all occurred since 2001, with the exception of 1998.

Climate Change is turning up the heat on Idaho's Chinook salmon, as well as many other species worldwide, and is becoming an ever-increasing threat to their survival. To spawn successfully, salmon need the right combination of temperatures and stream flows at exactly the right time of the year. Unfortunately, warmer temperatures, considered to be an aspect of Climate Change, have been affecting the quantity and timing of snowmelt and runoff in Idaho mountains in the springtime.

Reduced snowfall and loss of snowpack from higher temperatures cause reduced river flows, making it more difficult for salmon to migrate to the sea in a natural timeframe. Similarly, slower river flows make it more difficult for salmon returning from the ocean to their spawning grounds.

Higher than normal temperatures can also trigger runoff from snowmelt to happen too quickly. This can cause flooding that can flush eggs and young salmon from their nests, making them easy targets for predators, or smothering them with silt from erosion caused by the flooding. Less snow and earlier snowmelt can also lead to drought, which makes wildfires more common. Large wildfires in the United States currently burn more than twice as many acres each year as they did in the 70s, and the average fire season now lasts 2 1/2 months longer. Wildfires destroy trees and roots that normally create shade and thus colder water in the Columbia River Basin salmon habitat, so wildfires contribute to warmer rivers and streams, as well.

Climate Change also causes warming air temperatures to create warmer rivers and streams. Salmon require cold water in order to thrive. 42-60° F is the best range of temperature for Chinook. Anything above this can be fatal for them. In 2015, for example, more than 100 migrating Chinook died in the John Day River, a tributary of the Columbia River, when water temperatures reached the mid 70s°F in a heat wave. Salmon fall prey to disease-causing bacteria and fungus and become stressed at temperatures above 68° F and stop migrating at or over 70° F. According to the EPA, studies show that by 2100 one third of current habitat for salmon in the Northwest will most likely be too warm for these species to tolerate if we do not act on Climate Change now.

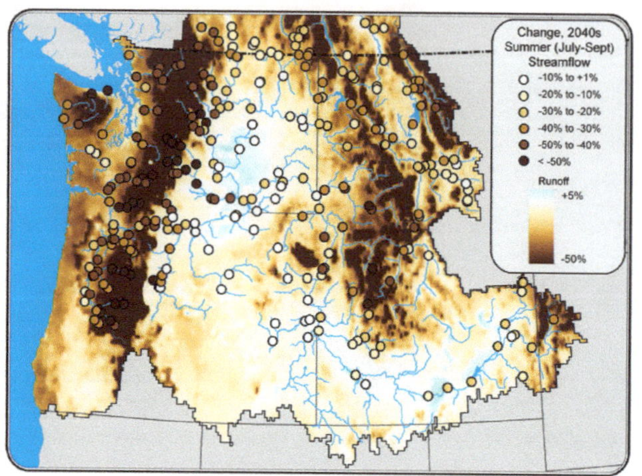

Natural surface water availability during late summer is projected to decline across most of the Northwest. This map shows expected changes in local runoff (shading) and streamflow (colored circles) for the 2040s (compared to the period 1915 to 2006), assuming that heat-trapping greenhouse gases will be reduced in the future. Credit: EPA

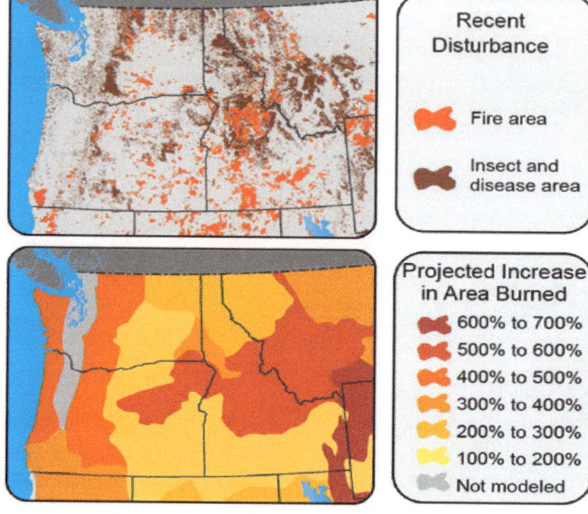

Under hotter, drier conditions, insects and fire can have large cumulative impacts on forests. This is expected to be the dominant driver of forest change in the near future. The top map shows areas burned between 1984 and 2008 or affected by insects or disease between 1997 and 2008. The bottom map indicates the expected increase in area burned resulting from a 2.2° F. warming in average temperature. Credit: EPA

Another climate related issue that Chinook face when in the ocean is "ocean acidification." Too much carbon dioxide (CO_2) and human activity is making oceans more acidic, which decreases oxygen levels in the water, harms coral reefs, and dissolves the shells of tiny mollusks that are a food source for salmon.

In late 2013, a so-called "blob" of warm water in the Pacific Ocean had reduced salmon productivity by as much as one tenth. Scientists are increasingly concerned about this becoming a regularity. In 2019, the Intergovernmental Panel on Climate Change determined that the ocean has taken up more than 90% of the excess heat in the climate systems, warning of the long-term impacts this may have on marine life and fish populations, especially in coastal areas.

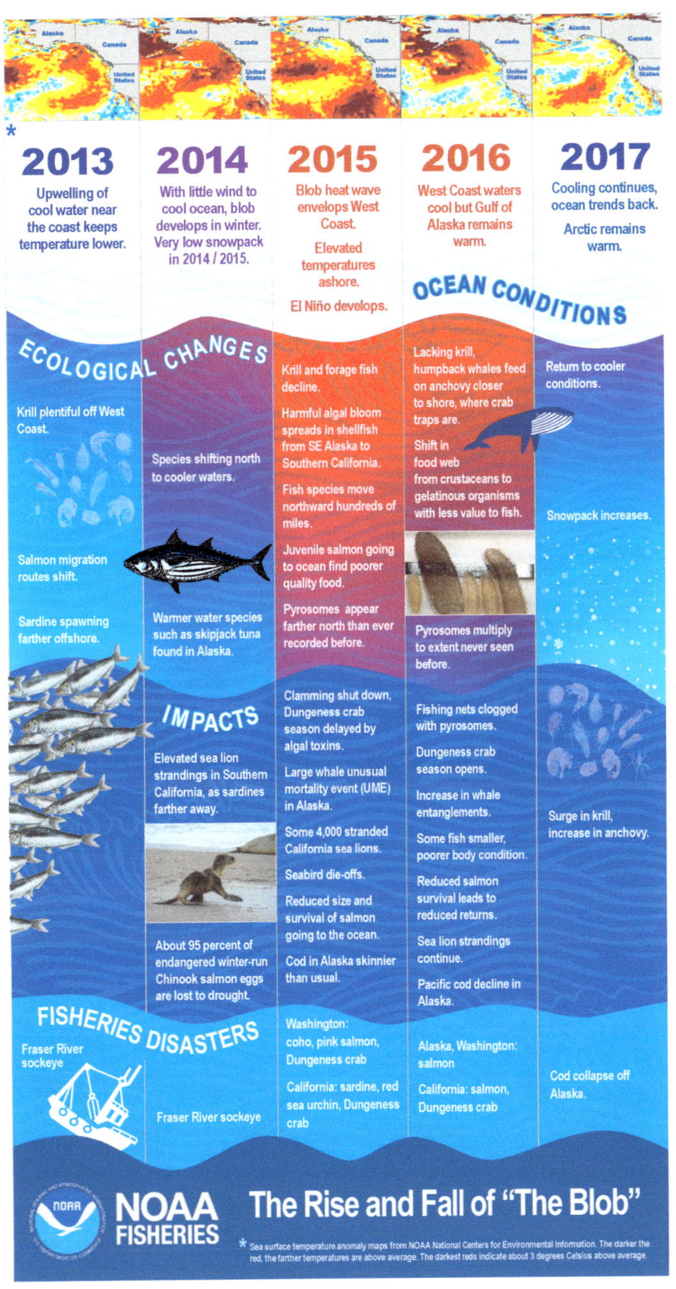

"Source: National Oceanic and Atmospheric Administration (NOAA)
Some climate scientists think that warming oceans will become "the new normal".

Water Pollution & Sediments

Human activities in industry and agriculture can feed us, provide lumber for giving us shelter, give us the materials to make anything from roads to jewelry, provide jobs, and enhance our lives in many ways. Unfortunately, some of these activities, if not carried out in an environmentally healthy way, can have damaging or lethal effects on the habitat of salmon and other organisms. Sediment from logging or mining roads pose a threat to streams where salmon eggs need to hatch. The sediment can suffocate eggs in a redd and decrease dissolved oxygen levels in the stream. This issue is improving as some newer regulations prevent these roads from being placed near breeding habitat of sensitive fish.

Another form of sediment pollution can come from agricultural lands near streams. If cattle are allowed to graze next to and drink from streams where salmon live, too much sediment can affect water quality. In addition, **excrement**, or digestive waste, either from direct access to streams or from runoff from irrigation ditches or canals, can pollute the stream by increasing bacteria and algae loads.

Some ranching communities have graciously helped with this problem. Ranchers who own land along the Lemhi River in Eastern Idaho, where salmon spawn, serve as a model of how ranchers and conservationists can come together for a common cause. They have collaborated with state and federal agencies and conservation groups to restore and improve Chinook habitat on the Lemhi for years. "The important thing that we'd like people to understand is, ranchers are equally or more concerned about all of the species here than anyone," Carl Lufkin, manager of the Little Eight Mile Ranch near Leadore, Idaho says. "And I think the results that we're getting here show that we're getting that done."

Livestock can be a problem for water quality for salmon. Credit USDA

Biodiversity Matters

As you read earlier, marine biologists say that if Snake River Chinook salmon populations continue to decline, so will predators who depend on them. **Biodiversity**, the interconnected nature and interdependence of a variety of species in a food chain, means that all living things depend on other living things to survive. For example, producers like plants need the carbon dioxide that we exhale, like we need the oxygen they produce though photosynthesis, as well as the food they provide. Other consumers, such as foxes, need to eat rodents as a main source of their diet. If rodents suddenly disappeared, foxes' lives would be drastically changed, and they would have to adapt or face possible extinction. The problem with adaptation is that it can take thousands of years.

Idaho's Snake River Chinook salmon are a keystone species because they are critical to the survival of the Southern Resident Killer Whales (SRKWs) in the Puget Sound inlet in Washington. In fact, Snake River Chinook are their main prey in winter and spring.

Chinook are an important food source for these killer whales because Chinook are the largest of the Pacific Salmon and killer whales prefer large fish to meet their energy needs. To give you an idea of the magnitude of this relationship, the endangered "J Pod" of orcas, which consists of around 75 use residents in Puget Sound, must together consume about 723 Chinook salmon a day in the winter and spring, at least to sustain their needs. This means that they need at least a quarter of a million Chinook a year just to survive! A healthy, recovered population of this pod would need even more than that to flourish, perhaps as much as 30%, according to Rob Williams of Oceans Initiative.

And it's not just the numbers. Idaho's Chinook salmon are getting smaller in size, especially within the last fifteen years. The ocean is where they typically gain their large size as they age, but there has been a recent trend in which Idaho's Chinook are spending fewer years at sea, which may have to do with the Pacific Ocean becoming more acidic and warmer. The declining size of Chinook salmon makes a difference, as the smaller Idaho's Chinook get, the more of these fish killer whales will

Southern Resident killer whales use echolocation clicks to detect orientation of a Chinook's swim bladder.

Southern Resident Killer Whale (Orca) mother and her calf. Credit: NOAA

need to eat to be properly nourished.

With Idaho's Chinook populations declining, the J Pod killer whales of Puget Sound are in distress and on a downward spiral headed toward extinction. In fact, they were placed on the Endangered Species list in 2005.

In 2018, a 20-year-old mother orca of the "J Pod", named Tahlequah, or J57, carried her dead calf for 17 days and 1,000 miles in what appeared to be a display or grief. It is thought that the calf, who was just a day old and still milk dependent, died of starvation, as Tahlequah was most likely malnourished due to the low numbers of Chinook available. Captured by the news, Tahlequah's display of her dead calf created a greater awareness of this problem, which is a good start.

Critical Solutions

"Since Idaho's salmon and steelhead were listed as threatened or endangered under the Endangered Species Act in 1991, more than $16 billion taxpayer and [hydroelectric power] ratepayer dollars have been spent on failing federal recovery strategies," says Zack Waterman of the Sierra Club. "The irony that these funds originate from the hydropower system which threatens anadromous fish survival is not lost on those working closely on recovery efforts," writes Laurie Sammis, in the article "The Great Migration" for Sun Valley Magazine. This is a lot of money to be spent on strategies that haven't worked for decades of efforts, and people want to see these salmon make a come-back. One such person who'd like to see this happen is Mike Simpson, an Idaho Congressman since 1999. In a video he launched in 2021, called "What If?", he agrees that too much has been spent on salmon recovery without enough to show for it, as Idaho salmon are still in crisis. He says, "It would be a tragedy if future generations looked back and wished our generation of leaders and stakeholders would have taken the time to explore this opportunity to develop our own Northwest solution to protect stakeholders and save salmon."

Solutions to the Dam Problem

The Bonneville Power Company and Army Corps of Engineers in charge of building the dams in the Columbia River Basin knew that salmon would need help getting around these dams, but clearly, they didn't anticipate all the problems salmon would face from these unnatural structures. Even with ideas like building fish ladders and bypass systems, navigating these dams is still proving to cause too much stress and the possible extinction of salmon in Idaho. Biologists and engineers continue to investigate ways to restore salmon populations, but solutions that attempt to work around the dams have not been successful so far in the last 60-70 years of problem-solving this issue and billions of dollars spent.

Hatcheries

One solution that has been around since the 1800s is the building of state, tribal, or federal fish hatcheries to compensate for the losses of wild fish caused by dams and for harvesting fish for food. An entire federally-funded network of hatcheries in Washington, Oregon, and Idaho was created just for the Lower Snake River dams' impact because of how lethal they were expected to be for salmon and steelhead.

Chinook born in hatcheries are not wild fish but are descendants of wild fish. When hatchery fish return to the river at the hatchery to spawn, they are blocked from swimming any further and collected to be "manually spawned." Salmon are quickly and humanely killed (all Pacific salmon die shortly after spawning in the wild) by hatchery workers so that they can harvest the females' eggs and the males' milt to fertilize them. After the eggs are fertilized, they are placed into rearing trays in water to incubate. When the eggs hatch and later become fry, they are moved to water-filled raceway ponds, or canals, to grow into smolts and are then released into the river, or trucked to a specific section of river, to migrate to the ocean and begin the process all over again.

Hatchery fish, which are able to be harvested, have taken some pressure off of wild salmon and steelhead, which are not allowed to be harvested by fishermen. Hatcheries usually remove the adipose fin from hatchery smolts so that they cannot be confused with wild salmon. Smolts at hatcheries are also "PIT" tagged (Passive Integrated Transponders) with a microchip for identification. When a salmon crosses a dam on the Snake or Columbia Rivers, the dam's microchip-readers process the tag information so biologists can keep track of how many salmon or steelhead are successful.

Hatcheries have done good work to save salmon, and many people believe that they are needed despite some of the problems that may result,

Chinook smolts with red PIT Tags Credit: Nez Perce Tribal Fisheries

Fall Chinook egg rearing trays at the Nez Perce Tribal Hatchery. Credit: Wayne Penney, Nez Perce Tribal Fisheries

although there are some biologists and conservationists who disagree. One concern is that a few salmon may sometimes stray from the hatchery waters and successfully spawn in the wild. If they mate with wild salmon, biologists think that they may introduce weaker genetics to the wild gene pool with very few wild fish left. Currently, about 80% of the adult salmon returning to the Columbia River Basin are the product of hatcheries, and all of Idaho's wild salmon populations are listed as threatened or endangered under the Endangered Species Act.

Increased Spill

A recent strategy put forth to save fish from being killed or injured by dam turbines is to increase the spill-flow over dams when salmon are migrating to the sea. This takes a hit on the hydroelectricity output and in some cases is not allowed with state water laws, but in 2017, a U.S. District Court judge ordered dam managers in the Columbia River Basin to develop a plan to spill more water over the dams when salmon and steelhead need it because other methods had not been successful. Some say it has been more success-ful than the other bypass systems, but there are also other problems. When large quantities of water drop hundreds of feet, it creates too many dissolved gases, called "nitrogen supersaturation" when it collects at the bottom of the spillway. This can cause "gas bubble dis-ease" in fish, in which bubbles of gas form in their gills, eyes, and other parts of their bodies, creating lesions. This can be fatal for juvenile salmon.

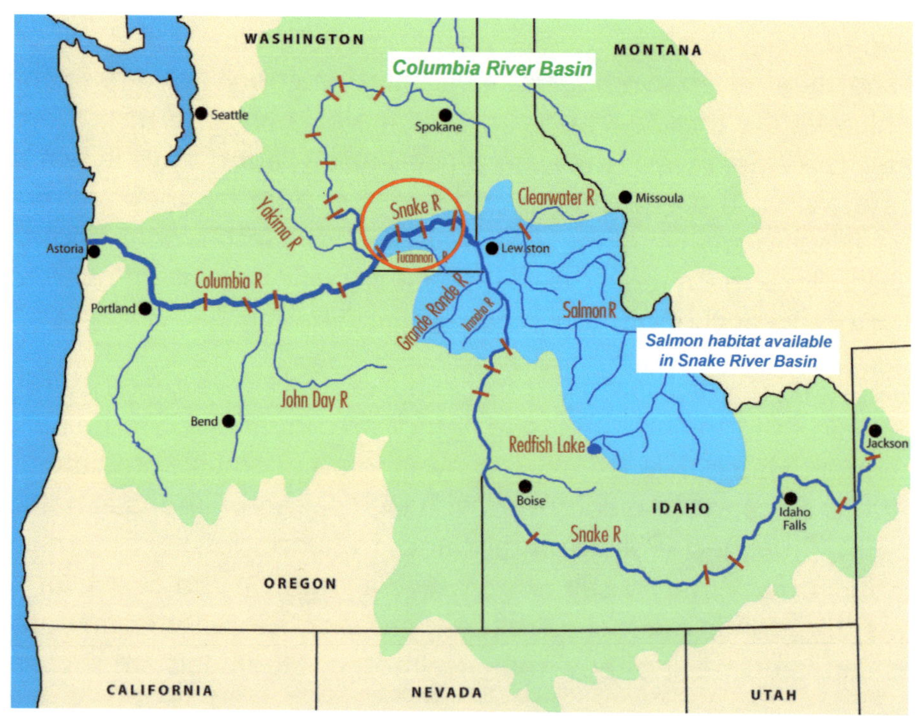

The red marks in the circle represent the four dams that biologists say could possibly restore around 70% of the salmon and steelhead population if "breached" or removed. Credit: Bert Bowler

Breach the 4 Lower Snake River Dams

An increasingly popular solution that scientists say could restore salmon populations to healthy levels is to breach, or take down, the four Lower Snake River dams. These were among the last of the dams built in the Columbia River Basin. Before these four dams were constructed, Idaho's migrating salmon still had to navigate the four larger dams on the Columbia River, but their populations remained stable. Once these four lower Snake River dams were in the picture, salmon numbers began to drop dramatically, illustrating that these new dams are the biggest problem.

Breaching these four dams on the Snake River has been a controversial issue. The dams, which are managed by a public utility company, Bonneville Power, were built over 45 years ago to create low-cost power to millions of people, to help irrigate farms, and to ship large quantities of grains from Idaho to the West Coast for farmers. The case to keep these dams in place to provide cheap electricity is weakening lately as solar and wind energy and natural gas have become equally, if not more, low-cost sources of power. In addition, shipping of grains to the coast has declined by 70% since dams were built, as farmers have begun to use railways and trucks more often for transportation of grains from Idaho.

Salmon advocacy groups, as well as many economists, say that the cost of maintaining these aging dams, which are not generating much money, makes it hard to justify the spending, especially when Idaho is losing so many salmon. Daniel Malarkey, author of The Sightline Study, says, "These dams have reached the end of their useful lives. They are going to take a lot more additional investment if they're going to be operated for another 30 years, and when you factor in their true costs, including their effect on fish, they are no longer the low-cost resource. We have low-carbon and zero-carbon alternatives that don't make fish go extinct."

Conservationists say that breaching a dam is not as complicated as it may seem. It can be done by removing the earthen section of the dam, which would create a channel for fish to swim around the dam, actually leaving the concrete structures in place. Some say, even if it isn't done for the fish, there is justification for eliminating these dams purely on a cost savings basis.

Some good news is that in February of 2021, Congressman Simpson of Idaho wrote a comprehensive proposal to improve the infrastructure of the Pacific Northwest, including a plan to remove the four aging lower Snake River dams to help migrating fish, while considering and compensating those who have benefited from these four dams. Although this is just a proposal and doesn't call for the dams' removal until 2030, it has given salmon advocates some hope that progress is coming.

Some Climate Change Solutions

"As I see it, the scientific lesson is clear: If the habitat is available and healthy, the salmon know how to recover…"

- Jim Lichatowich, author of Salmon Without Rivers.

Like you read earlier, Climate Change is the result of burning too many fossil fuels, such as oil, gas, and coal, which overloads the atmosphere with carbon dioxide (CO_2). Climate Change is a global crisis, not just a problem in the United States. The good news is that the ocean, trees and other plants, and even the soil, can absorb large amounts of carbon dioxide from the atmosphere. The not-so-good news is that the ocean is absorbing too much CO_2, and too many trees are being cut down, especially in the Amazon Rainforest. The soil that once could hold a large amount of carbon is often being overly disturbed with pesticides, too much tilling, planting of monocrops, over-grazing, and excessive use of fertilizers, which all allow the carbon dioxide to burn off into the atmosphere again. There are a multitude of things that humans can do to offset Climate Change, but there are five main actions that society can focus on to begin to make an impact!

What's Up With Refrigerators?

Every refrigerator, supermarket case, and air conditioner contain chemicals that absorb and release heat so that it is possible to keep food, buildings, and cars cool. Refrigerants like chlorofluorocarbons [klor-o-FLOR-o-kar-binz] (CFCs), which release into the air, have luckily been discontinued after a gaping hole in the atmosphere's ozone layer was discovered. The ozone layer (made up of O_3 molecules) protects living things on Earth from the harmful ultraviolet radiation from the sun. Three decades later, the ozone layer is beginning to heal, showing us that our actions do matter. However, other refrigerants, like hydrochlorofluorocarbons [hy-dro-KLOR-o-FLOR-o-kar-binz] (HCFCs) continue to harm the planet and need to be phased out. This will take many years to change out old refrigerators and air conditioners, and 90% of the HCFC emissions happen during disposal, so that needs to be considered.

One solution is to carefully manage the destruction of these refrigerants as they are phased out, purify them for reuse, or turn them into chemicals that do not cause warming in the atmosphere. Climate advocates say that we need to start this management with the biggest offenders of HCFCs use: large stores like Walmart. According to Green America, these actions could reduce possibly 89 gigatons of **greenhouse gases** by 2050.

Another solution is to embrace new technologies that are beginning to provide potential for less harmful refrigerants. Natural refrigerants like "hydrocarbon refrigerants," made popular by Ben & Jerry's Ice Cream freezers, are not yet widely available in the U.S. yet have been used for over 20 years in other parts of the world. This type of refrigeration that uses naturally occurring hydrocarbons has no ozone-depleting properties, is considered to be climate-friendly, and is more cost-effective than older refrigerants. Hopefully, these will become more available in the U.S. soon.

Wind & Solar Energy

The Earth has many renewable resources that can be used in place of energy sources that harm the environment like fossil fuels, or sources that harm migrating fish, like hydroelectric dams. One renewable source that has been gaining a lot of attention in the last decade is wind power. Current technologies are making it easier to make wind energy more effective and it is a clean alternative to using fossil fuels. Of course, the key is finding windy places. According to Green America, there are four states alone in the U.S. that could generate enough energy from wind power to meet electricity demand from coast to coast: Texas, Iowa, Kansas, and Oklahoma. Today, over 300,000 wind turbines supply 4% of global electricity, and the addition of more of these could provide many new jobs. Wind energy is becoming one of the least expensive sources of electricity capacity. According to the U.S. Office of Energy Efficiency & Renewable Energy, "By 2050, wind energy could avoid the emission of 12.3 gigatons of greenhouse gases."

Wind Energy Credit: Energy.gov

Solar energy is another option for cutting greenhouse gases that affect Climate Change. The sun is probably the greatest source of clean, renewable energy. Trees must have time to regrow to be used as fuel, and although wind is a valuable renewable resource, it may be somewhat less consistent than solar rays. One problem is that it has been expensive to convert a home's energy system to solar. The good news is that the U.S. Department of Energy aims to reduce the price of solar energy by 50% by 2030.

As reported by a study in 2017 by the National Renewable Energy Laboratory, generating 35% of electricity from solar and wind energy in the Western U.S. would reduce CO_2 emissions by 25-45%. They also claim that solar and wind energy now provide the cheapest source of power for 67% of the world.

Reduction in Food Waste

Hunger is a problem for nearly 800 million people in the world, yet 1/3 of food raised doesn't reach people's plates. In areas where income is low, this occurs due to lack of refrigeration or proper storage facilities, ineffective packaging, too much heat and humidity, or bad roads.

In higher income areas, food waste is more of the problem of choice. Store owners and customers reject food that is bruised, discolored, or simply doesn't look perfect. Also, people simply order too much food, overestimate how much food to buy for the week, and toss out food that they have forgotten about in the back of the refrigerator.

The food we waste each year translates into an equivalent of nearly 4 gigatons of CO_2 into the atmosphere and wastes water, energy, land, labor, and money along the way. In lower income areas, improving transportation, processing, and storage is needed, as well as utilizing local and small farms to reduce the fuel it takes to ship food from far away. Some conservationists say that, in higher income areas, having less food in stores may prevent people from buying in excess when not needed, as well as making expiration dates more about when food becomes unsafe versus when food should taste best.

Climate scientists say that if we, as a global society, reduce waste in this way by 2050, we could possibly see 70 gigatons of reduced CO_2.

The Global Carbon Cycle

330 Gt CO2 340 Gt CO2 440 Gt CO2 450 Gt CO2 40 Gt CO2

Over time, the ocean and land have continued to absorb about half of all carbon dioxide emissions, even as those emissions have risen dramatically in recent decades. It remains unclear if carbon absorption will continue at this rate. Credit: NASA/JPL-Caltech

Restore the Rainforests

On average, one acre of new forest can **sequester** about 2.5 tons of carbon each year. Tropical rainforests can store up to 250 billion metric tons of carbon in the trees alone, saving that carbon from reaching the atmosphere. Unfortunately, these rainforests are being chopped down or burned to make room for livestock or farms at an alarming rate in recent decades, and much of this is done illegally. Poverty and deforestation are definitely linked. In poor, rural areas, like in the Amazon rainforests, people sell their trees for a quick profit or burn them down to make room for growing food because they desperately need money. Rainforests now cover just 5% of Earth's land compared to a previous 12%. The loss of rainforests today is responsible for at least 18% of greenhouse gas emissions caused by human activity. These forests absorb and hold lots of carbon, so the good news is that the United Nations Strategic Plan for forests 2017-2030 goal is to increase the world's forested areas by 3% by 2030, while reducing poverty, which leads to this destruction of rainforests.

Fuel-Efficient Vehicles

Highway vehicles release about 1.7 billion tons of greenhouse gases into the atmosphere each year, mostly in the form of carbon dioxide, contributing to Climate Change each year. Each gallon of gasoline burned creates 20 pounds of greenhouse gases, which equals to roughly 6-9 tons of greenhouse gases each year for a typical vehicle!

One study estimated that we could cut fuel consumption by 10% by switching to our highest miles-per-gallon (mpg) cars available when feasible. In 2017 there were 65 million multi-vehicle households in the United States. Each of these households, according to the Environmental Protection Agency (EPA), drives an average of 28,000 miles per year. If each one reduced fuel consumption by 10% a year by switching to a more gas efficient vehicle, we would need 10 billion fewer gallons of gasoline each year. Not only would this save drivers money, but this could prevent almost 100 million metric tons of tailpipe CO_2 emissions!

An even better way to cut CO_2 emmisions is to drive a hybrid or electric vehicle. Many automobile companies are currently designing hybrid or fully electric models of their cars and trucks, which help cut down on carbon emmisions. In fact, General Motors announced in January 2021 that the company would phase out gas-powered vehicles completely and replace them with cars and trucks that have zero tailpipe-emmisions by 2035. This will be good news for the planet, yet more needs to be done sooner. In the meantime, we need to be driving vehicles that are the most gas efficient.

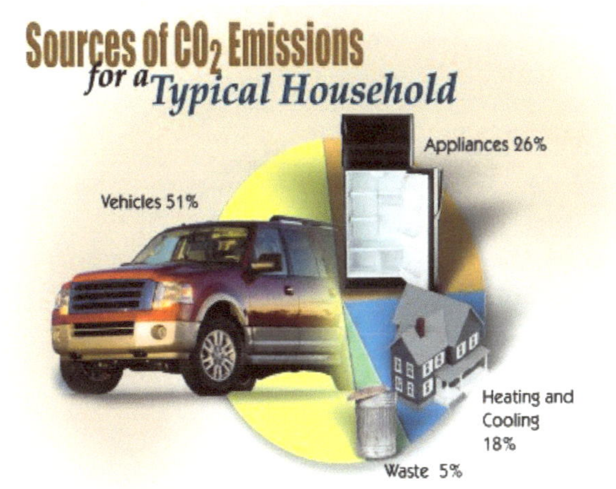

Source: Fueleconomy.gov

Save our Chinook! What You Can Do

Some things you can do to help protect and restore salmon habitat include:

- Conserve electricity by turning off lights and unplugging appliances when not in use.

- Conserve water by turning the faucet off while brushing your teeth and by taking shorter showers.

- Use energy-efficient appliances.

- Use energy-efficient light bulbs.

- Weatherize your home to keep the heat in during winter and to keep cool in the summer.

- Reduce pesticide and fertilizer use and make sure they don't runoff into waterways.

Snake River "Flotilla" gathering to support a free-flowing Snake River. Lewiston, Idaho, 2019. Credit: Snake River Waterkeeper

- Advocate for solar and wind energy instead of hydroelectric power.

- Ride your bike instead of driving to nearby places.

- Recycle & use less plastic. Use paper bags instead of plastic and use reusable containers instead of plastic baggies. Plastic breaks down into methane and ethylene, increasing the rate of Climate Change and harms many aquatic animals.

- Pick up litter, as it could wash into storm drains and streams.

- Don't waste food. Only buy what you need and actually eat it.

- Adopt a nearby creek or river and volunteer to clean it up and keep it healthy.

- Make your voice heard! You are never too small in size or number to make a difference! Write letters to the lawmakers representing Idaho and tell them that you want them to do more to save Idaho salmon and steelhead before it's too late. Go to this website to learn how to contact them.

https://legislature.idaho.gov/legislators/contactlegislators/

Glossary

adipose fin: a small fleshy fin located on the back of a salmon between the dorsal and caudal fins. Research suggests that this fin may serve as a sensor to help the caudal fin, or tail, maneuver better in turbulent water.

alevin: a newly hatched salmon or trout that still has a yolk sac attached. Alevin use the yolk sac for food during the first month of life.

anal fin: an unpaired fin located on the underside of a fish near the tail and used for balance and maneuvering.

apex predator: a predator at the top of a food chain that is not preyed upon naturally by any other animal.

anadromous: fish who are born in freshwater, then migrate to the ocean as juveniles where they grow into adults before migrating back into freshwater to spawn.

anatomy: science concerned with the bodily structure of humans, animals, and other living organisms, especially as revealed by dissection and the separation of parts.

biodiversity: a relationship of the interconnectedness of a variety of species, land and seascapes, and the cultural links to the places where we live—whether right where we are or in distant lands—important to our wellbeing as they all play a role in maintaining a diverse and healthy planet.

breach: the act of breaking or rupturing. The term "dam breaching" means to take down a dam that is no longer useful or causing problems for migrating fish.

broodstock: in salmon hatcheries, brood stock are a group of mature salmon used in hatcheries for breeding purposes and to sustain their population. In Chinook hatcheries, around 120,000 adult fish are needed to return from the ocean to sustain the population.

caudal fin: the tail fin, which is used to propel the fish. It is also used by female salmon to dig a redd, or nest, in a stream, in which to deposit her eggs.

consumer: a living organism that eats other organisms from a different population because they cannot make their own food like producers do. Example: a Chinook salmon (consumer) eats plankton (producer), but also eats insects, crustaceans, and smaller fish (other consumers).

dam: a dam is any retaining structure, built across a river to stop or regulate its flow, and to raise the water level. Behind the dam, water accumulates to form a lake or reservoir.

decomposer: an organism, like a bacterium, fungus, or insect (like maggot), that breaks down the cells of dead plants and animals into simpler substances to transfer energy and nutrients throughout an ecosystem.

dense: particles or components that are closely compacted together. Example: saltwater is more dense than freshwater so the saltwater will sink below the freshwater.

dorsal fin: an unpaired fin on the back of a fish used to balance and stabilize the fish from rolling over.

eggs: eggs are the very first stage in a salmon's life. They contain a developing embryo and food store, which are surrounded by a soft membrane for protection.

erosion: a process in which earthen materials are carried away and transported by forces such as wind or water, often destroying streambanks and filling streams with silt.

estuary: a body of water where fresh and saltwater mix at an area where a river meets the sea. Migrating salmon spend time in estuaries to transition from living in freshwater to living in saltwater.

excrement: waste system, discharged from the digestive system of an animal.

extinction: the state or process of a species dying out in a specific area, which they once naturally inhabited, or in totality.

food chain: a series of organisms each dependent on the next as a source of food in a hierarchy.

freshwater: a body of water having a low salt concentration (usually less than 1%) versus the ocean, which has a high salt concentration. Examples of freshwater are streams, rivers, lakes, reservoirs, ponds, wetlands, groundwater, and glaciers. Only about 3% of earth's water is freshwater.

fry: the stage of a salmon between alevin and parr, when the yolk sac is almost entirely gone so they must find food for themselves. At this stage, they leave the protection of the gravel nest and begin swimming to feed on plankton.

genus: a class or group of living things. In biology, it means two or more species that share unique body structures or other characteristics are considered to be closely related and are placed together in a genus. The genus is the first part of the scientific name of a species and is always written with a capital letter and in italics. The "genus" of Pacific Chinook salmon is *Oncorhynchus* and the "species" is *tshawytscha*.

gills: are the body part that help a fish breathe underwater. In fish and other aquatic creatures, their gills are equivalent to our lungs, except the oxygen they take in is dissolved in the water.

greenhouse gases: are gases like carbon dioxide and methane that, when released into the Earth's atmosphere, trap heat.

habitat: the natural home or environment of an animal, plant, or other organism.

hatchery: a place where the hatching of fish or poultry eggs is artificially controlled.

homestead: is land claimed by a settler or squatter, or person laying claim to the land without ownership (as in settlers illegally homesteading on First Nation People's original land and reservations.

hydroelectric dam: a dam producing hydropower, which is electricity produced from generators thrusted by turbines that convert the potential energy of falling or fast moving water from a reservoir into mechanical energy.

indigenous: originating from or occurring naturally in a region. The Nez Perce Tribe is **indigenous** to the Northwest area of the United States. They already lived in this area before Europeans arrived.

juvenile: juvenile Chinook are typically defined as being two years of age or less.

keystone species: a species on which other species in an ecosystem largely depend, such that if it were removed, the ecosystem would change drastically in a negative way.

kype: the hook on the lower jaw of a mature male salmon.

lock: a system that allows boats and ships to bypass a dam by going into a lock, which is a large chamber in the water with gates at each end that can open and close when a valve is opened. The water from the opened lock flows into the next body of water, raising or lowering the boat automatically. The boats continue into the next lock, and so on, until they reach the end of the dam and lock system.

lateral line: Low frequency sounds are detected in the water through the lateral line, a system of fluid-filled sacks with hair-like sensory features that are open to the water through pores along the side of the fish. Salmon do not have ears, so the lateral line allows them to detect movement of other fish and predators in the water.

macroinvertebrate: an animal that doesn't have a backbone and is big enough to see without a microscope. Examples are insects and worms.

magnetoreceptor: is an organ or magnetite in living tissue that detects magnetic fields, particularly the Earth's magnetic field. Some form of magnetic sense is found in a wide range of animals, including insects, fishes, amphibians, reptiles, birds, and mammals.

migrate: animals, birds, or fish move from one region or habitat to another according to the seasons.

milt: a substance produced by male spawners, which contains sperm to fertilize a female's eggs.

natal: relating to the place or time of one's birth.

olfactory: relating to a sense of smell.

paired fins: are located on each side of a fish, primarily in guiding the forward course of movement and in maneuvering within the water column.

parr: the stage of a salmon between the fry and smolt stages, having dark, rounded, vertical markings on their sides.

photosynthesis: the process by which green plants and some other organisms use sunlight, carbon dioxide and water to make their own food. Photosynthesis in plants generally involves the green pigment chlorophyll and generates oxygen as a byproduct.

predator: an organism or animal that exists by **preying** on, or hunting and eating, other organisms or animals.

prey: an organism hunted or captured as food for another organism or animal, called a **predator**.

producer: organism on the food chain that can produce its own energy (from the sun) and nutrients. Example: plankton is a producer.

redd: a spawning nest made by a salmon or trout where a female deposits eggs and a male deposits sperm to fertilize the eggs.

riffle: a rocky, swift, and shallow section of a stream, providing good spawning habitat.

scales: scales protect fish from predators, viruses, fungi, and parasites and help a fish glide through the water.

semelparous: organisms reproducing or producing offspring only once in a lifetime. Examples are biennial plants,

many insects, and a few vertebrates.

sequester: to capture carbon dioxide from the atmosphere and store it.

smolt: the stage between parr and juvenile in the salmon life cycle, in which the young salmon is getting ready to migrate to the ocean.

smoltification: a complex series of physiological changes in which a young salmon adapts from living in freshwater to seawater. In fresh water, the salmon's body is saltier than the water in which it swims. To work properly, the body needs salt, so it tries to keep the salt in. Some escapes, but the salmon gets enough from the food it eats to make up for the loss. In the ocean, the water is saltier than the salmon's body needs to be, so it must try to keep the salt out and the water in. When salmon swim in the ocean, the saltwater draws water out of the fish's cells. Salmon adapt by drinking sea water to replace the water their cells lose.

spawn: fish spawn (or mate) to produce offspring. The female deposits eggs in a gravel nest (a redd) that she built by swishing her tail to loosen and deposit gravel, and at the same time, the male fish deposits sperm over the eggs to fertilize them. This is called spawning.

species: a group of individuals that breed,

or mate, together to produce off-spring. Individuals of a species cannot breed with other species to produce fertile offspring. The species is the second part of a scientific name for a species and is always written with a lowercase letter and in italics. The "species" name for Pacific Chinook salmon is *tshawytscha.*

substrate: (of a river or stream) the bottom layer underneath a body of water containing pebbles, gravel, sand or silt, plants, and organisms.

tide: the periodic inflow and outflow of the waters of the ocean and its inlets, produced by the attraction of the moon and sun, and occurring about every 12 hours.

tepee: is a short wall of hides that is strung around tree poles on the inside of the tepee cover. It makes the tepee like an "envelope house" where the cold air from outside enters at the bottom of the tepee cover, goes up several feet between the cover and the liner, then enters the inside already pre-warmed. It creates a ventilation system that ensures that the tepee is cool in summer, warm in winter, and not nearly as smoky or wet as a tent. Some say it is an engineering marvel.

treaty: a legally binding agreement between nations.

tributary: a river or stream flowing into a larger river or lake.

vertebrate: an animal of a large group or species that has a backbone, including mammals, birds, fishes, and reptiles.

Index